Undergraduate Lecture Notes in Physics

Undergraduate Lecture Notes in Physics (ULNP) publishes authoritative texts covering topics throughout pure and applied physics. Each title in the series is suitable as a basis for undergraduate instruction, typically containing practice problems, worked examples, chapter summaries, and suggestions for further reading.

ULNP titles must provide at least one of the following:

- An exceptionally clear and concise treatment of a standard undergraduate subject.
- A solid undergraduate-level introduction to a graduate, advanced, or non-standard subject.
- A novel perspective or an unusual approach to teaching a subject.

ULNP especially encourages new, original, and idiosyncratic approaches to physics teaching at the undergraduate level.

The purpose of ULNP is to provide intriguing, absorbing books that will continue to be the reader's preferred reference throughout their academic career.

Series Editors

Neil Ashby
University of Colorado, Boulder, CO, USA

William Brantley
Department of Physics, Furman University, Greenville, SC, USA

Matthew Deady
Physics Program, Bard College, Annandale-on-Hudson, NY, USA

Michael Fowler
Department of Physics, University of Virginia, Charlottesville, VA, USA

Morten Hjorth-Jensen
Department of Physics, University of Oslo, Oslo, Norway

Michael Inglis
Department of Physical Sciences, SUNY Suffolk County Community College, Selden, NY, USA

More information about this series at http://www.springer.com/series/8917

Rudolf P. Huebener

Conductors, Semiconductors, Superconductors

An Introduction to Solid-State Physics

Third Edition

 Springer

Rudolf P. Huebener
Physikalisches Institut
University of Tübingen
Tübingen, Germany

ISSN 2192-4791   ISSN 2192-4805   (electronic)
Undergraduate Lecture Notes in Physics
ISBN 978-3-030-31419-4   ISBN 978-3-030-31420-0   (eBook)
https://doi.org/10.1007/978-3-030-31420-0

This book was originally published in German by Spektrum Verlag, 2013. Original title: "Leiter, Halbleiter, Supraleiter: Eine Einführung in die Festkörperphysik".

This Springer imprint is published by the registered company Springer Nature Switzerland AG
The registered company address is: Gewerbestrasse 11, 6330 Cham, Switzerland

Preface to the Third Edition

The third edition is a reprint of the second edition with a series of minor extensions. Because of the strong recent experimental and theoretical research activities in the field of topological materials, Sect. 11.6 has been extended considerably.

The author is grateful to Zachary Evenson from the Springer editorial office for his effective support.

Tübingen, Germany   Rudolf P. Huebener

Preface to the Second Edition

The second edition is an extended reprint of the first edition. In many chapters, a more detailed mathematical treatment of the various subjects is given. A section dealing with magnetic impurities and the Kondo effect has been added.

The author is grateful to Silvia Haindl and to Vladimir Kresin for detailed suggestions about expanding the book.

Tübingen   Rudolf P. Huebener

Preface to the First Edition

Only a few scientific-technical developments from the last century have affected our lives in such a powerful way as the spectacular advances in our knowledge of the electronic properties of solids. Many of the present achievements are intimately connected with these advances. To name only a few: the transistor and its extreme miniaturization in microelectronics, the electronic processing of data and highly developed and powerful computers, the mobile telephone and satellite communication, television and entertainment electronics, as well as numerous instruments and systems of medical technology.

In the final analysis, the theater of all these events of dramatic progress is the world of electrons in crystals, where the (quantized) vibrations of the crystal lattice continuously demonstrate their influence. The revolutionary advances in knowledge are due to many individual people. Frequently a true paradigm change has been necessary in order to arrange and order the new perceptions properly. Hence, it is not surprising that, as a rule, the pioneers of these new ideas initially had to overcome great difficulties and rejection, before the new concepts slowly gained acceptance. Also, in certain cases, highly focused research in large industrial laboratories turned out to be the key to success. This is impressively illustrated in particular by the invention of the transistor in the American Bell Laboratories.

This book represents an updated and strongly extended edition of the book published by the same author nearly 10 years ago with the title *Electrons in Action*. In particular, the physical contents were pointed out more clearly by mathematically formulating the fundamentals. The book is aiming at students of the natural sciences, and in particular of physics and materials science, as well as at engineers, as an introduction to solid-state physics. It may serve as a motivating pre-stage and companion of the established and very detailed textbooks.

In addition to the physical contents, the book treats the important role played by many famous and often still very young scientists. The fundamental developments are supplemented by describing their scientific and historic environment.

Marius Orlowski, Virginia Polytechnic Institute, provided important advice.

Tübingen   Rudolf P. Huebener
May 2014

Contents

Mathematical Symbols

a	Distance between the neighboring atoms or building blocks of the crystal lattice
e	Electric elementary charge
f	Spring constant
$\mathbf{f}_L$	Lorentz force
h	Hour
h	Planck's constant
$\hbar$	$h/2\pi$
j	Electrical current density
j_s	Supercurrent density
k	Wave number $= 2\pi/\lambda$
$\mathbf{k}$	Wave vector $= k_X + k_Y + k_Z$
k_Z	Component of the wave vector in z-direction
$\mathbf{k}_F$	Fermi wave vector
k_B	Boltzmann's constant
ℓ	Mean free path
m	m_e = Electron mass
m_h	Mass of the holes
m_c	Cyclotron mass of the electrons
n	Electron concentration
n_s	Density of the superconducting electrons
p	Hole concentration
$\mathbf{r}$	Lattice vector
$\mathbf{r}$	Electron coordinate
t	Time
v_D	Drift velocity
v_F	Fermi velocity
z	Figure of merit
mA	Milliampère
mg	Milligram
nm	Nanometer $= 10^{-9}$ Meter

μm	Micrometer = 10^{-6} Meter
eV	Electron volt (energy unit)
A	Vector potential
B	Magnetic flux density
C_V	Specific heat at constant volume
D_2	Density of states in the two-dimensional case
D_e	Density of states per volume
E	Energy
F	Free energy
E	Electric field
G	Vector in the reciprocal lattice
G	Density of the Gibbs free energy
G_o	Quantized unit of the electrical conductance
GeV	Giga-electron volt = 10^9 eV
GHz	Giga-Hertz = 10^9 per second
H	Magnetic field
H_C	Critical magnetic field of a superconductor
H_c	Coercive field
H_{C1}	Lower critical magnetic field of a superconductor
H_{C2}	Upper critical magnetic field of a superconductor
I	Electrical current
I_r	Recombination current
I_g	Generation current
I_C	Critical electrical current density of a superconductor
J	Exchange integral
K	Kelvin
K	Wave vector of the phonons
L	Lorenz number
L	Orbital angular momentum
M	Magnetization
N	Number of the crystal atoms
R	Electrical resistance
R_H	Hall constant
S	Seebeck coefficient
S	Spin angular momentum
S_m	Mixing entropy
T	Temperature
T_C	Critical temperature of a superconductor
T_D	Debye temperature = $h\nu_D/k_B$
T_{CU}	Curie temperature
THz	Tera-Hertz = 10^{12} per second
U	Total vibrational energy of a crystal
U	Inner energy
V	Electrical voltage
V	Volume

α	Polarizability
α	Madelung constant
δ	Scattering angle
κ	Heat conductivity
κ_e	Heat conductivity of the electrons
κ_G	Heat conductivity of the crystal lattice
θ	Debye temperature
ε	Electron energy
ε_F	Fermi energy
λ	Wavelength
λ_m	Magnetic penetration depth of superconductors
μ	Magnetic moment
μ_o	Permeability of vacuum
μ_B	Bohr magneton
η	Damping coefficient of the motion of magnetic flux quanta
ω	Angular frequency
ω_B	Bloch frequency
ω_c	Cyclotron angular frequency
ω_D	Debye angular frequency
ω_L	Larmor frequency
ν	Frequency
ν_c	$\omega_c/2\pi$ = Cyclotron frequency
ν_D	$\omega_D/2\pi$ = Debye frequency
ν_E	Einstein frequency
χ	Magnetic susceptibility
χ_p	Paramagnetic susceptibility
τ	Mean collision time
ξ	Coherence length of a superconductor
$\boldsymbol{\rho}$	Lattice vector
ρ_f	Flux-flow resistivity
σ	Electrical conductivity
φ	Phase of a wave function
φ_o	Magnetic flux quantum
φ_0	Atomic wave function
ψ	Wave function of the superconducting Cooper pairs
Π	Peltier coefficient

Chapter 1
Spectacular Advances

Abstract During the second half of the last century solid state physics and materials science experienced a great advance and established itself as an important and independent new field. In addition to X-ray diffraction, new analytical tools such as neutron diffraction, electron microscopy, different versions of mechanical scanning techniques, and scanning electron and laser microscopy became available. Material fatigue, radiation damage, and the preparation of single crystals developed into important subjects. The invention of the transistor represented perhaps the ultimate highlight.

During the second half of the last century the physics of solids has experienced a tremendous growth, for which many important basic steps had already been prepared during the first half of the century. An early decisive impulse for these developments came from the discovery of X-rays in 1895 in Würzburg, Germany, by Wilhelm Conrad Röntgen. Soon afterwards, this discovery lead to the first observation of X-ray diffraction in crystals by Max von Laue in 1912 in Munich. William Henry Bragg, Professor in Leeds, England, together with his son William Lawrence Bragg at the early age of only 22 years, then started the systematic analysis of crystal structures by means of X-ray diffraction.

Today, research dealing with the physics of solids has an impressively wide scope, if for no other reason than the fact that solids are always needed to fabricate useful or nice things, in contrast to the totally different role of liquids and gases. The exact knowledge of the physical properties of the materials that we use today becomes more and more important the further we advance in the field of high technology. The large effort of research and development within the area of solid state physics becomes obvious if one looks at the program books for the relevant annual meetings of, say, the German Physical Society (DPG) or the American Physical Society (APS) which recently contained up to more than 2000 pages. (Today, these programs of the meetings are distributed mostly electronically).

Often, the technological applications provide the key motivation for strong basic research in solid state physics. We illustrate this by the following two examples. On January 10, 1954, an English passenger airplane of the Comet type broke apart at

© Springer Nature Switzerland AG 2019
R. P. Huebener, *Conductors, Semiconductors, Superconductors*, Undergraduate Lecture Notes in Physics, https://doi.org/10.1007/978-3-030-31420-0_1

8200 m altitude in the Mediterranean near the isle of Elba without any prior warning and crashed into the sea. With only 3681 flight hours, the plane was relatively new. The search for the cause of the accident turned out to be extremely difficult, even though people worked feverishly to clarify the cause of the terrible crash. Since the cause of the accident continued to remain unknown, it was finally concluded that the crash must have been the result of an unfortunate combination of several bad effects. Hence, on March 23, 1954, the grounding order for all airplanes of the same type, immediately issued on the day of the accident, was lifted again. Prior to this, a total of 62 modifications had been introduced in all Comet airplanes in operation or under construction. In this way it was hoped to exclude any possible cause of the accident (Fig. 1.1). Then a completely unexpected dramatic event happened. On April 8, i.e., only 16 days following the resumption of the regular flight operation, another Comet airplane with only 2704 h of flight operation crashed into the Mediterranean near Naples. Again, at a high altitude of 10 000 m this time, the plane suddenly apparently broke apart. Now the situation became extremely serious. The causes had to be found at the highest level, and all available means had to be utilized. After analysis of the many different possibilities, problems related to what is now called material fatigue, in particular associated with the wings, came to the center of attention. As a consequence, the complete fuselage of an airplane was dumped into a huge tank filled with water in order to expose it to changing, and in particular to cyclical, mechanical loads. In this way, it was found that, after some time, fatigue effects appeared on the wings. However, the fatigue problems on the fuselage itself were much more severe. Finally, the evidence became clear that the mechanical load during testing caused cracks in the fuselage, and that all the cracks originated at the rectangular corners of the cabin windows. The causes of both plane crashes had been found. However, this event also put to an abrupt end the British leading role in air traffic. (Today a large piece of a side of one

Fig. 1.1 Comet jet-aircraft beginning a test flight after the crash of a plane in the Mediterranean near the isle of Elba. (*Photo* ullstein bild)

of the two crashed aircrafts, recovered from the Mediterranean, is on display in the Science Museum in London).

Because of these dramatic developments, intensive research activities were begun at the same time at many places. Until then only little was known about the phenomenon of material fatigue, its effect on the mechanical properties of materials, and the mechanisms leading to the development of microcracks.

In this context of the material fatigue experienced sixty years ago with the Comet passenger airplane, it is interesting to note that the recent development of the largest passenger airplane, which has ever been constructed, the Airbus A 380, included an extensive and careful mechanical material fatigue testing procedure by means of hydraulic systems, as a critical step. Starting in 2005 the complete A 380 airplane, consisting of the whole fuselage and the wings, has been exposed for 26 months to mechanical loads varying with time and simulating a total number of 47500 flight cycles (take-off and landing). This testing load program corresponds to the 25 year lifetime of the A 380 airplane.

As a second example, we recall the possible difficulties expected more than 60 years ago during the operation of the inner components of the first nuclear reactors. At that time hardly anything was known about the behavior of, say, graphite when it is utilized for slowing down the neutrons which are emitted during nuclear fission within the reactor. Would it be possible that during their irradiation with the highly energetic neutrons the carbon atoms of the graphite lattice could be ejected out of their regular lattice sites, eventually leading to an energetically highly excited material, releasing abruptly its stored excess energy in an explosion like dynamite? Such problems concerned the scientists involved in the early reactor experiments. The American scientist Eugene Paul Wigner, originally from Hungary (later a Nobel laureate and famous for his theoretical work on mathematical group theory and symmetry principles and their role in atomic, nuclear, and elementary particle physics) was one of the first who theoretically analyzed the physical properties of lattice defects and radiation damage in crystals. At that time, a young co-worker of Wigner, Frederick Seitz, performed the first theoretical calculations on this subject (Fig. 1.2). Both scientists introduced the concept of the "Wigner-Seitz cell" into solid state physics. Following these initial steps, the field of structural lattice defects in crystals has developed into an important subfield of solid state physics, being investigated today in many laboratories. In 1940 Frederick Seitz also published the first general textbook on solid state physics: "The Modern Theory of Solids".

An enormously important development took place with respect to microelectronics. Here the physics of solids has resulted in a total paradigm change in electronic technology. It was Mervin Kelly, one of the top-level managers of the famous American Bell Laboratories in Murray Hill in the Federal State of New Jersey, who realized at the end of the Second World War that the old mechanical relays and the evacuated amplifying tube made from glass had to be replaced by something better. To Kelly a highly promising candidate appeared to be the crystal, if it had suitable electric conduction properties. Therefore, at the Bell Laboratories a special group of scientists was organized, which was supposed to explore the

Fig. 1.2 Eugene P. Wigner (left photo: Deutsches Museum) and Frederick Seitz (right: private photo)

electric conduction properties of solids. At the center of everyone's attention then stood the semiconductor crystals of germanium and silicon. Already, relatively soon afterwards, an extremely momentous event had been the invention of the transistor by John Bardeen, Walter Brattain, and William Shockley. On December 23, 1947, they demonstrated the transistor for the first time to the directors of their company. Subsequently, as a new electronic device, the transistor underwent intensive further development and improvement. Without a doubt, this invention represented the start of the modern age of digital electronics.

These big advances in the field of solid state physics, of course, were accompanied by similar advances in instrumental techniques and methods. Here we must mention the exploration of the regime of very low temperatures. In 1908 the Dutch scientist Heike Kamerlingh Onnes in Leiden achieved for the first time the liquefaction of the noble gas helium. With this success the low-temperature range down to 4 K (−269 °C) became accessible. In this context the most spectacular event was the subsequent discovery of superconductivity by Kamerlingh Onnes in 1911. Until the 1930s, the number of laboratories equipped to perform experiments with liquid helium worldwide could be counted on the fingers of one hand. In contrast, today about 1000 helium liquefiers are operating worldwide (Fig. 1.3). Today the largest liquefaction facility is operated at the particle accelerator *Large Hadron Collider* (LHC) in Geneva. There exist eight liquefiers each having a liquefaction rate of 3600 L/h, i.e., with a total rate of 28 800 L/h. Worldwide this corresponds to about 40% of the inventory of large liquefaction facilities for helium.

Eventually, the available experimental regime was extended to lower and lower temperatures. In particular, we mention a technique relying on the elementary

Fig. 1.3 Modern plant for liquefying the noble gas helium. On the left we see the controls and the cold box of the liquefier, on the right the storage vessel for liquid helium. (*Photo* Linde AG)

atomic magnets of a paramagnetic substance. This technique consists of the following sequence of steps. Initially, a paramagnetic salt pill is precooled to about 1 K, in order to reduce considerably its content of thermal energy. Subsequently, the elementary magnets in the salt pill are all oriented in one direction by a strong magnetic field, and simultaneously the heat of magnetization is removed, being deposited in the environment. In the next step, the salt pill is thermally decoupled from its environment. Then the magnetic field is turned off. Now the pill is thermally isolated, and the directional disorder of the elementary magnets gradually reappears. As a necessary consequence, the temperature of the salt pill drops at the same time. In this way low temperatures of only a few thousandth Kelvin can be reached. This method of "adiabatic demagnetization" was proposed in 1926 by the Dutch scientist Peter Debye and in 1927 by the American William Francis Giauque. In 1933 the method was demonstrated experimentally for the first time. The extended application of this principle to the elementary magnets of the atomic nuclei had already been proposed in 1934 by the Dutch scientist Cornelis Jacobus Gorter and in 1935 by Nicholas Kurti and Franz Eugen Simon from Oxford. The cooling effect due to this nuclear demagnetization was experimentally realized for the first time in 1956. Using this technique, extremely low temperatures down to one millionth Kelvin or lower could be reached. However, at such low temperatures it becomes more and more difficult to establish thermal equilibrium between the different components of the solid, namely the electrons and their elementary magnets, the lattice vibrations, and the elementary magnets of the atomic nuclei.

Because of their Jewish origin, Nicholas Kurti and Franz Eugen Simon had to leave Germany in 1933 when Hitler took over the government. Earlier, both had worked first in Berlin and then at the Technical University in Breslau (today Wroclaw), and the English scientist Frederick Alexander Lindemann (later Viscount Cherwell) had arranged for a position for both of them at the Clarendon Laboratory in Oxford, England. As director of the Clarendon Laboratory Lindemann has done exactly the same at the time also for the two brothers Fritz and Heinz London, and for Kurt Mendelssohn. After they had left Germany, during subsequent years, all these people distinguished themselves by outstanding contributions to physics at low temperatures, and Oxford gained a top position in this field.

An apparatus often used today for reaching temperatures much below 1 K is the mixing cryostat (Fig. 1.4). In this cryostat the two isotopes of the noble gas helium, which differ only by the number of neutrons in their atomic nuclei (^{3}He with a single neutron and ^{4}He with two neutrons), are pumped through several stages of heat exchangers, such that within the mixing chamber located at the coldest end of the instrument an almost pure liquid ^{3}He phase is collected directly above a liquid mixed phase of ^{3}He and ^{4}He. For this technique to operate, the starting temperature must already have been lowered to 1 K by precooling. During operation, ^{3}He atoms from the upper concentrated phase are dissolved continuously in the lower, much more diluted phase. In many ways this scheme resembles a regular evaporation process, in which the upper phase corresponds to the liquid and the lower phase to the vapor. As the final result, a continuous cooling of liquid helium is achieved.

Fig. 1.4 Mixing cryostat for cooling down to temperatures well below 1 K. The coldest end with the mixing chamber is located at the bottom. On the top one can see the flange for mounting into the cryogenic container, which can also be evacuated. (*Photo* Oxford)

With this apparatus the attached sample to be studied can also be cooled continuously. The lowest temperatures which can be reached are a few thousandth Kelvin. The principle of the mixing cryostat was proposed for the first time in 1951 by Heinz London. The first prototype was operated in 1965. Together with his brother Fritz London, Heinz London also had proposed an early theory of superconductivity.

In addition to the continuing improvements in experimental instruments and to the refinements of measuring techniques, sample preparation and the development of materials also saw much progress. Here an important step was the production of single crystals with extremely high purity. It was such ultra-pure single crystals which allowed the exact determination of many physical properties of materials and the achievement of a theoretical understanding based on these data (Fig. 1.5). The growing of large single crystals starts by dipping a little seed crystal under an inert gas atmosphere into the melt of the same material and then pulling it out again at a slow and well regulated speed. In this way, during solidification of the melt, the exact atomic order of the seed crystal will be reproduced. Record sizes of such cylindrical single crystals up to more than one meter in height and nearly half a

Fig. 1.5 Silicon single crystal. (*Photo* Wacker Chemie AG)

meter in diameter have been achieved. The concentration of atomic impurities in such a crystal can be reduced further by means of the "zone melting process". During this process the total cross-section of a short length of the crystal is heated up to the melting temperature by means of, say, eddy current heating, while this heating zone is slowly moved from one end of the crystal to the other. In the resulting temperature gradient the atomic impurities are carried along to one end of the crystal. If necessary, this process can be repeated several times. The impurity concentration of silicon single crystals, routinely achieved today in the semicon- ductor industry, amounts to only about a single impurity atom within one billion silicon atoms.

The spectacular advances in our physical understanding of the microscopic properties of solids were closely coupled to the progress of the instruments and methods available for the analysis of materials. In addition to the investigation of the structure of crystals by means of X-ray diffraction already mentioned, starting in the 1950s the diffraction of neutrons was also utilized more and more for clarifying crystal structures. For this purpose, special nuclear reactors built only for research purposes served as neutron sources. As an example we mention the egg-shaped research reactor ("Atomei") built in the 1960s at the Technical University of Munich in Garching, Germany (Fig. 1.6). In some sense as a training ground, this reactor then turned out to become the point of origin for the much larger research reactor of the German-French Laue-Langevin Institute in Grenoble. In 2004 the "Atomei" was replaced by the new Research Reactor FRM II (research neutron source Heinz Maier-Leibnitz) in Garching. Similar construction projects for research reactors existed also in other countries with a highly developed industry.

Fig. 1.6 Egg-shaped research reactor ("Atomei") in Garching near Munich. In the building on the left the new research reactor FRM II, completed in 2004, is located. (*Photo* Albert Scharger/TU Munich)

In the same way as often happens with new ideas, the invention of the electron microscope initially had to withstand great difficulties and rejection. It all began with two Ph.D. students, namely Ernst Ruska and Bodo von Borries, who had joined the group of Max Knoll at the Chair of High-Voltage Engineering and Electric Plants at the Technical University of Berlin during December 1928 and April 1929, respectively. Here, at first both worked on the improvement of the cathode ray oscilloscope. Because of the experience gained, before long they developed the idea that beams of fast electrons can be used for generating a magnified image in a new type of microscope. On March 17, 1932 Ernst Ruska and Bodo von Borries submitted their first and basic patents on the future electron microscope. However, a few large hurdles still remained to be overcome. "Why do we need electron microscopes, since we have light microscopes?" was the question that people were asking. However, soon both young scientists had a breakthrough. The Company Siemens & Halske in Berlin agreed to pick up the idea and prepared employment contracts for Bodo von Borries and Ernst Ruska. On December 7, 1937 the first electron microscope built by Siemens was demonstrated to the Company directors (Fig. 1.7).

Fig. 1.7 Siemens electron microscope, a precursor of the Siemens Elmiskop 1, marketed in the 1950s. (*Photo* TU Berlin)

After only three years of development, in terms of its spatial resolution the electron microscope had outpaced the light microscope. Starting in 1939, an initial series of the Supermicroscope ("Übermikroskop"), as it was called at the time, was offered for sale by Siemens.

Again, the underlying basic concept of this microscope is the quantum-mechanical wave character of elementary particles, which had been proposed for the first time by the French Louis de Broglie in his dissertation in 1924. The direct experimental proof of the wave nature of electrons was provided subsequently in 1927 by the two Americans Clinton Joseph Davisson and Lester Germer of the Bell Telephone Laboratories who showed that electrons are diffracted by the atomic lattice of crystals. During imaging based on the diffraction of waves, the spatial resolution is always limited by the wavelength. The shorter the wavelength, the correspondingly smaller are the structures that can be spatially resolved. The wavelength of the beam electrons is inversely proportional to the square-root of the accelerating voltage. At an electric voltage of 10 000 volts we have a wave length of $\lambda = 1.2 \times 10^{-2}$ nm (nm = nanometer = 10^{-9} m). On the other hand, the wave length of visible light is much larger, $\lambda = 400–800$ nm, and the achieved spatial resolution is correspondingly much weaker.

Already in the 1950s, electron microscopy had celebrated a big success, along with many other successes, by imaging the structural defects in the crystal lattice, as discussed above, and by clarifying the phenomenon of material fatigue. In the latter case the "crystal dislocations" play a central role. They were observed directly for the first time in 1956 at the Battelle Institute in Geneva in stainless steel and at the Cavendish Laboratory in Cambridge in aluminum. Eventually, electron microscopes were built for ever increasing accelerating voltages. Today we have instruments with an accelerating voltage of one million volts (Fig. 1.8).

For the analysis of materials, beams of fast electrons have also been utilized in another important instrument: the scanning electron microscope. For this, pioneering research was done again in the 1930s by Max Knoll at the Technical University in Berlin, mentioned before, and by Manfred von Ardenne in his Laboratory in Berlin-Lichterfelde. An electron beam collimated down to an extremely small diameter of only 1–10 nm is scanned over the surface of the object to be investigated. Simultaneously, a suitable signal induced by the electron beam in the sample is recorded as a function of the spatial beam coordinates on the sample surface within the scanning window. Correct electronic signal processing then yields a two-dimensional image of the object. To generate the response signal one can use several effects. For example, the emission of secondary electrons due to the beam irradiation is quite often used. However, the beam-induced local change of a sample property such as the electric resistivity can also provide the signal for the image. Today, the signal based on the change in electric resistivity is often utilized for imaging structures in thin layers of semiconductors or superconductors. In the case of superconductors, spatially resolved images relating to their superconductivity can be obtained if the sample is cooled to sufficiently low temperatures during scanning with the electron beam.

Fig. 1.8 Modern electron microscope with an accelerating voltage of one million volts. (*Photo* A. Tonomura)

Recently, the scanning principle for imaging was extended also to light beams. However, a necessary prerequisite for this was the availability of laser beams with their extremely narrow collimation. Today, laser scanning microscopes are widely used in many fields.

An important milestone during the advances of the methods for the analysis of materials has been the construction of the first scanning tunneling microscope by Gerd Binnig and Heinrich Rohrer of the IBM Research Laboratory in Rüschlikon near Zürich, Swizerland. Their first patent application dealing with the scanning tunneling microscope was submitted in January 1979. In their instrument the surface to be investigated is mechanically scanned with a tiny metal tip. Using piezoelectric actuators, the metal tip can be moved in three dimensions with extremely high sensitivity. During the scanning process the sample surface is approached by the tip as close as about 1 nm. Simultaneously, the quantum-mechanical electric tunneling current is measured running between the tip and the sample surface, if an electric voltage is applied, even though a metallic contact between both does not exist. (The explanation of the effect of quantum mechanical tunneling had been one of the early major successes of the new theory of quantum mechanics). Because of the strong exponential dependence of the tunneling current on the distance between the tip and the sample surface, one can achieve that the tunneling current is limited only by a few or even the last single atom sticking out of the tip. In this way, today one routinely obtains atomic resolution in the lateral direction with this technique (Fig. 1.9). Very recently, even subatomic structures of silicon atoms due to the different electron orbitals, have been observed in the images (Fig. 1.10).

Fig. 1.9 Scanning tunneling microscope. The instrument is mounted on a flange for operation in ultra-high vacuum. (*Photo* OMICRON Nano Technology)

Fig. 1.10 Image of the individual atoms of a section of 5 nm × 5 nm area on the surface of a silicon crystal, generated by means of atomic force microscopy. On the silicon atoms we can see a subatomic structure resulting from the electron orbitals. (*Photo* F. J. Giessibl)

Fig. 1.11 The picture shows a ring of iron atoms placed on a copper surface. In this way an artificial coral reef consisting of 48 iron atoms has been created on an atomic scale. The circular lines appearing within the ring are due the density of the electrons existing within the ring. (*Photo* Almaden Research Center, 2000)

Soon after the invention of the scanning tunneling microscope, the mechanical scanning principle was extended to several other types of interaction between the probing tip and the sample surface. In particular, we mention the atomic force and the magnetic force microscopes. In the first case, the mechanical force between the probing tip and the sample surface is utilized. The second case is based on a magnetic tip probing the magnetic sample properties. In recent years special research effort has been concentrated on the extension of the techniques we have discussed to very low temperatures and to the presence of high magnetic fields. Today, ease of operation is emphasized by the construction of the instruments.

Finally, we point out that most of the techniques for material analysis discussed above are restricted to the sample surface and its immediate neighborhood (Fig. 1.11).

In many cases the developments we have outlined were accompanied by the award of the Nobel Prize for Physics and in some cases for Chemistry to the people involved. In order to illustrate this, in the Appendix we have listed all Nobel Laureates who have a close relationship with the physics of solids.

Chapter 2
Well Ordered Lattice Structures in Crystals

Abstract The lattice structure of crystals is characterized by specific symmetry properties. Translation symmetry yields the 14 Bravais lattices. Rotation, reflection at a mirror plane, and inversion at a point result in the 32 crystallographic point groups. The diffraction of X-rays by a crystal, initiated in 1912 by Max von Laue, represented the first experimental proof of the regular lattice structure of a crystal. The elements of diffraction theory, including the reciprocal lattice and Brillouin zones, are explained. The chapter ends with a discussion of quasi-crystals and the different types of bonding.

Crystals have always generated a particular fascination, because of the rich variety of their colors and shapes. While the systematic exploration of nature became increasingly important ever since the 17th century, at the same time the science of rocks and minerals developed into an independent branch and a collecting point for the many different individual observations. The amateur rock collectors and the mineralogists hiking with their tools through the mountains and hills in the early days must be looked upon as important forerunners of the modern scientific exploration into the properties of solids. The basic geometric crystallographic concepts for describing the large variety of observations also originated within this field of mineralogy.

In terms of physics, the most important property of crystals is their perfect lattice structure with the regular periodic repetition of exactly the same elementary building blocks in all three spatial dimensions. The elementary building blocks can be atoms or molecules, the latter consisting either of only a few or very many individual atoms. For example, the elementary building blocks of protein crystals contain up to 100 000 atoms. Because of their highly regular periodic lattice structure, crystals always possess a number of prominent symmetry properties. Of particular importance is the "translation symmetry" resulting from the regular periodic lattice configuration of the building blocks in all three spatial dimensions.

In a crystal the location of the building blocks of the lattice is described mathematically by the lattice vectors.

© Springer Nature Switzerland AG 2019   15
R. P. Huebener, *Conductors, Semiconductors,*
Superconductors, Undergraduate Lecture Notes in Physics,
https://doi.org/10.1007/978-3-030-31420-0_2

$$\mathbf{r} = n_1\mathbf{a} + n_2\mathbf{b} + n_3\mathbf{c} \tag{2.1}$$

Here, n_1, n_2, and n_3 are integers. $\mathbf{a}$, $\mathbf{b}$, and $\mathbf{c}$ are the three fundamental translation vectors. Here and in the following we denote vectors by bold symbols. The integer numbers n_1, n_2, n_3 yield the lattice points of the crystal. (In (2.1) we assume, that the origin is located at a lattice point). The translation vectors $\mathbf{a}$, $\mathbf{b}$, $\mathbf{c}$ generate the elementary cell (Fig. 2.1), which in turn builds up the crystal lattice by its spatially periodic repetition. Because of this condition of translation symmetry, the possible configurations of all three-dimensional crystal lattices are highly restricted. As was shown already in 1850 by the Frenchman Auguste Bravais, there are only a total of 14 fundamental types of crystal lattices, which are now referred to as "Bravais lattices" (Fig. 2.2). By selecting the lengths of the three vectors $\mathbf{a}$, $\mathbf{b}$, $\mathbf{c}$ of the elementary cell (lattice constants) and the three angles between them, at first one obtains seven fundamental types of crystal lattices. If additional lattice points exist at special locations within the elementary cell (in the center of the elementary cell or in the middle of the external surfaces), one obtains a total of 14 translation lattices.

In general, the crystal structure is more complex than that of one of the 14 Bravais lattices. However, the crystal lattice is exactly replicated by means of a specific symmetry operation. In addition to translation, the following fundamental symmetry operations are important: rotation, reflection at a mirror plane, and inversion at a point. In the case of rotation one distinguishes how often the crystal lattice is exactly reproduced during a complete rotation by 2π. Hence, there exist single-, two-, three-, four-, and sixfold rotational axes, corresponding to a rotation by 2π, $2\pi/2$, $2\pi/3$, $2\pi/4$, and $2\pi/6$, respectively. The combination of rotation, reflection at a mirror plane, and inversion specifies one of the 32 crystallographic point groups. By addition of the translation, one of the 230 space groups is obtained, characterizing the crystal structure. Here, mathematical group theory has provided an important input.

Johannes Kepler, who was born in 1571 in the Swabian Free City Weil der Stadt near Stuttgart in Württemberg and who later studied at the University of Tübingen, is generally known because of his three famous Kepler's laws of astronomy. However, among many things he also was concerned with the question, how can space be regularly and completely

Fig. 2.1 Elementary cell defined by the translation vectors $\mathbf{a}$, $\mathbf{b}$, and $\mathbf{c}$

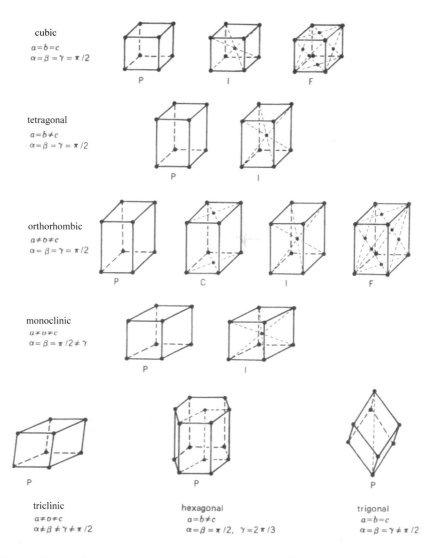

cubic
$a=b=c$
$\alpha=\beta=\gamma=\pi/2$

P   I   F

tetragonal
$a=b\neq c$
$\alpha=\beta=\gamma=\pi/2$

P   I

orthorhombic
$a\neq b\neq c$
$\alpha=\beta=\gamma=\pi/2$

P   C   I   F

monoclinic
$a\neq b\neq c$
$\alpha=\beta=\pi/2\neq\gamma$

P   I

P   P   P

triclinic
$a\neq b\neq c$
$\alpha\neq\beta\neq\gamma\neq\pi/2$

hexagonal
$a=b\neq c$
$\alpha=\beta=\pi/2, \ \gamma=2\pi/3$

trigonal
$a=b=c$
$\alpha=\beta=\gamma\neq\pi/2$

Fig. 2.2 The fourteen Bravais lattices representing all possibilities for the construction of a three-dimensional crystal lattice

filled with the same objects as building elements. So in the early 17th century, i.e., more than 200 years before the considerations of Auguste Bravais, he speculated on the question of why snowflakes always have six corners, but never five or seven. He showed how the close packing of spheres generates a six-corner pattern. This work of Kepler clearly represents an early significant contribution to geometrical crystallography.

Within the crystal lattice, there always exist specific planes along certain directions, which are closely and perfect periodically packed with atoms or with the elementary building blocks. On the outside of the crystal these planes then represent the extremely smooth and flat surface planes. This fact, together with the existing symmetry properties, is utilized extensively by the jewelry industry during the polishing of precious stones. In snow crystals the large variety of shapes is particularly impressive. Thomas Mann has well described this magnificent appearance of the snowflakes in his novel "The Magic Mountain" ("Der Zauberberg", here in an English translation):

> Brilliant clips, medals of decoration, jewelry stars, such as the most accurate jeweler cannot produce in a richer way and with more minute precision…., and among the myriads of magic little stars in their hardly visible, secret little splendor, not meant for the human eye, not a single one was equal to another.

The first rigorous experimental proof of the regular lattice structure of crystals was given in 1912 at the University of Munich. Soon after his great discovery of X-rays in Würzburg, Röntgen (Fig. 2.3) had left this location, since he had accepted an offer from the University of Munich. In Munich his group, together with the theoretical physicists at the Chair of Arnold Sommerfeld, concentrated on the problem of clarifying the nature of X-rays. The major question was whether X-rays are just electromagnetic waves such as visible light, but with a much shorter wavelength, or whether they are a new kind of particle radiation. Max von Laue (Fig. 2.4), a young member at the Chair of Sommerfeld, was thinking about diffraction experiments with X-rays. During the time when he was theoretically analyzing the diffraction of X-rays on lattices of points or of bars, he learned from a discussion with Paul Peter Ewald, a Ph.D. student of Sommerfeld, that crystals are

Fig. 2.3 Wilhelm Conrad Röntgen (*Photo* Deutsches Museum)

Fig. 2.4 Left: Max von Laue. Right: Set-up of the Laue experiment. On the left we note the X-ray tube and on the right the stage for mounting the crystal. (*Photos* Deutsches Museum)

likely to consist of a regular lattice arrangement of atoms. Von Laue noted immediately that crystals would be well suited for the diffraction of X-rays, as long as the distance between the atoms in the crystal and the wavelength of the X-rays had a similar magnitude. An initial estimate was encouraging. For the first experiments von Laue enlisted the help of Walter Friedrich, who had just been employed by Sommerfeld, and of Paul Knipping, a Ph.D. student of Röntgen, and success did not elude them. On June 8 and July 6, 1912 Sommerfeld was able to present the first X-ray diffraction images of a crystal to the Bavarian Academy of Science. This pioneering discovery meant the recognition of two important facts: X-rays are electromagnetic waves, and crystals consist of a three-dimensional regular lattice of atoms (or molecules).

For his discovery of X-rays, in 1901 Röntgen had received the first Nobel Prize in Physics. His letter to the Royal Bavarian State Ministry for Church and School Matters, in which Röntgen had asked for leave of absence in order to attend the award ceremony in Stockholm, is a highly interesting contemporary document, which we wish to quote at this point. On December 6, 1901 Röntgen wrote (here in English translation):

According to a confidential information of the R. Swedish Academy of Science the most respectful and devoted undersigned has been awarded the first Nobel Prize for the year 1901. The R. Swedish Academy is particularly keen that the Laureates personally receive the prize in Stockholm on the day of the award (Dec. 10). Since these prizes are of an exceptionally high value and are also highly honourable, the most respectful and devoted undersigned feels that he must follow, though not lightheartedly, the desire of the R. Swedish Academy and, therefore, he is asking for leave of absence for the coming week. Dr. W. C. Röntgen

2.1   Diffraction Theory

According to the theory of the diffraction of waves at a point lattice, during wave irradiation, from each lattice point there originates a wave which spherically propagates in all spatial directions (Fig. 2.5). We all know the similar wave propagation which takes place on the surface of water after we have thrown a stone into it. The spherical waves originating from the different lattice points of the crystal superimpose and become enhanced or extinguished. This is referred to as interference. We consider the (elastic) reflection of a wave at a series of parallel lattice planes (Fig. 2.6). We denote the distance between two neighboring lattice planes by a, and the angle between the planes and the direction of the incoming or the outgoing wave by θ. The difference of the distance covered by the wave upon the reflection at the two neighboring planes amounts to 2 a sinθ. In the case of

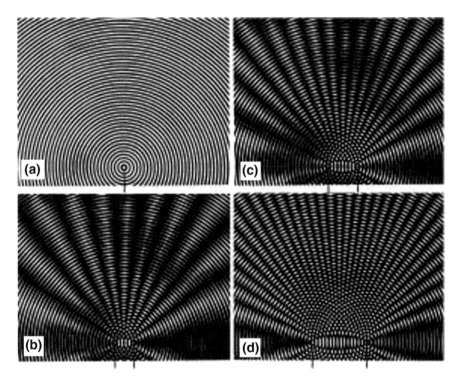

Fig. 2.5 **a** Propagation of a spherical wave originating from a point. The dark rings illustrate the peaks of the wave which follow each other within the spatial distance of one wavelength. The picture corresponds to a snapshot and shows the wave propagation within a plane as, for example, on a water surface. **b** Interference between two waves such as shown in **a**, originating from two different centers. From **b** to **d** the distance between the two centers increases. In specific directions the peaks and the valleys of the waves coincide, such that both waves annihilate each other. The positions at which the annihilation takes place, appear closer and closer as the distance between the two centers increases

constructive interference, at which the amplitudes of both waves exactly add to each other, this difference of the distance must be an integer multiple of the wavelength λ. Hence, we obtain the famous Bragg reflection law

$$2\,a\,\sin\theta = n\,\lambda \tag{2.2}$$

where n is an integer.

So far, we have restricted our discussion to a series of parallel lattice planes, which all have the same normal vector. However, in the case of a three-dimensional crystal we have to deal with two additional orientations of the lattice planes with the corresponding two directions of the normal vectors. For this, it is convenient, to start with a straight periodic arrangement of points forming a chain (Fig. 2.7). The spherical waves originating from all points are amplified, reaching a maximum intensity, if the propagation distance starting from two neighboring points differs exactly by one wavelength or by a multiple of one wavelength. On the other hand, if the difference amounts to half a wavelength or an uneven multiple of half a wavelength, complete extinction occurs. In this way we find that propagation directions for maximum or minimum intensity exist, and these are conically arranged around the straight line of points. The smaller the distance between the points, the larger is the opening angle of these cones. On the surface of an imagined sphere, having its center at the common tip of this family of cones, the propagation directions with maximum or minimum intensity form a series of circles. Next we extend our one-dimensional arrangements of points to a two-dimensional planar lattice. Now we must add a second family of cones which is arranged around the second newly-added straight line of points. On the surface of the imagined sphere, the propagation directions with maximum or minimum intensity yield a second series of circles. As a result, in this case the directions with maximum intensity are expected only for intersections between the corresponding circles originating from the two families of cones. However, such intersections always occur. Finally,

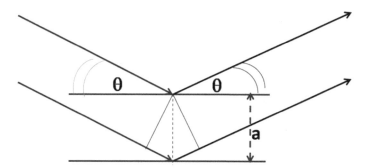

Fig. 2.6 Reflection of a wave at two neighboring lattice planes placed at the distance a. The difference of the distance covered by the wave upon the reflection at the two planes amounts to 2 a sinθ. θ is the angle between the lattice planes and the incoming and the outgoing wave, respectively

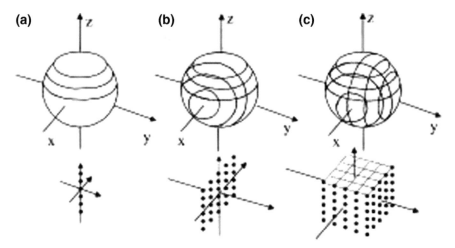

Fig. 2.7 Generation of the Laue diagram as the interference pattern due to the diffraction of X-rays at a three-dimensional point lattice. **a** For a one-dimensional chain of points, the directions of maximum or minimum intensity lie on cones arranged around the one-dimensional chain. On the surface of an imagined sphere, having its center at the common tip of this family of cones, the directions of maximum or minimum intensity yield a series of circles. **b** For a two-dimensional planar lattice of points we must add a second family of cones which is arranged around the direction of the second, newly-added straight line of points. Now a second series of circles appears on the surface of the imagined sphere. The points of intersection of these circles then yield the directions of maximum or minimum intensity. **c** Finally, for a three-dimensional point lattice we deal correspondingly with three families of cones. However, in general the resulting three series of circles on the surface of the imagined sphere no longer have common points of intersection, which mark the directions of maximum or minimum intensity. Now such common points of intersection only exist for special values of the wavelength or frequency of the X-rays

extending our discussion to a three-dimensional lattice of points, we have to deal with three families of cones. Again, it is the intersections between all three families of cones on the surface of the imagined sphere, which determine the propagation directions with maximum diffraction intensity. However, now the three series of circles on the imagined sphere do not in general have any common intersections (Fig. 2.7). In this case common intersections, marking the directions of high intensity of the diffracted waves, only exist as exceptions, i.e., only for specially selected values of the wavelength or frequency of the X-rays. For these selected wavelengths we have special distinct diffraction directions with high intensity, generating a characteristic pattern of points on the photographic film used for X-ray detection. This characteristic pattern is referred to as the "Laue diagram" (Fig. 2.8). However, this procedure only works if a whole frequency band of X-rays is available, from which the appropriate frequencies for the directions with high intensity are then automatically selected by the diffraction process.

The Bragg diffraction law expressed in (2.2) is named after the two English scientists William Henry Bragg and his son William Lawrence Bragg, whom we have previously mentioned. Immediately following the publication of the first Laue

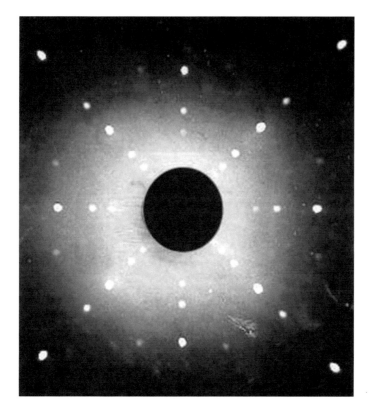

Fig. 2.8 Laue diagram of a cubic K_2SnCl_6 crystal obtained by X-ray diffraction (*Photo* J. Ihringer)

diagrams they have theoretically analyzed the underlying interference phenomena. The pattern of points on the Laue diagram is particularly useful for determining crystal symmetries.

2.2 Reciprocal Lattice, Brillouin Zones

Mathematically, the diffraction of a wave at a spatially periodic crystal lattice (periodic potential) can be treated conveniently by means of the concept of the reciprocal lattice. This concept is based on the abstract mathematical wave vector space or momentum space (Fourier space). It has been proposed by the American Josiah Willard Gibbs. A wave propagating along x-direction can be written as a complex function

$$F(x, t) = F_o e^{i(kx - \omega t)} \tag{2.3}$$

Here t denotes the time and ω the angular frequency. The wave number k is connected with the wavelength λ via the relation $k = 2\pi/\lambda$. In the three-dimensional case the function (2.3) can be generalized, and we obtain

$$F(\mathbf{r}, t) = F_o e^{i(\mathbf{k}\,\mathbf{r} - \omega t)} \tag{2.4}$$

where $\mathbf{r} = \mathbf{x} + \mathbf{y} + \mathbf{z}$ and $\mathbf{k} = \mathbf{k_x} + \mathbf{k_y} + \mathbf{k_z}$. Now the wave number k of the one-dimensional case is replaced by the wave vector $\mathbf{k}$ and its three components $\mathbf{k_x}$, $\mathbf{k_y}$, $\mathbf{k_z}$.

The reciprocal lattice is defined as follows:

$$\mathbf{G} = h_1 \mathbf{A} + h_2 \mathbf{B} + h_3 \mathbf{C} \tag{2.5}$$

where h_1, h_2, h_3 are integers. The fundamental vectors $\mathbf{A}$, $\mathbf{B}$, $\mathbf{C}$ are connected with the translation vectors $\mathbf{a}$, $\mathbf{b}$, $\mathbf{c}$ of the elementary cell (Fig. 2.2) and are defined as follows:

$$\mathbf{A} = 2\pi \frac{\mathbf{b} \times \mathbf{c}}{\mathbf{ab} \times \mathbf{c}}; \quad \mathbf{B} = 2\pi \frac{\mathbf{c} \times \mathbf{a}}{\mathbf{ab} \times \mathbf{c}}; \quad \mathbf{C} = 2\pi \frac{\mathbf{a} \times \mathbf{b}}{\mathbf{ab} \times \mathbf{c}} \tag{2.6}$$

We see that the vectors $\mathbf{A}$, $\mathbf{B}$, $\mathbf{C}$ of the reciprocal lattice are oriented perpendicularly to two fundamental axes of the crystal lattice, respectively. The convenience of the introduction of the reciprocal lattice becomes obvious, if one wants to express mathematically a function having exactly the periodicity of the crystal lattice (for example, the scattering potential of the electrons). Such a function is obtained in the form

$$U(\mathbf{r}) = \sum_{\mathbf{G}} u_{\mathbf{G}} e^{i\mathbf{G}\,\mathbf{r}} \tag{2.7}$$

where the summation extends over all vectors $\mathbf{G}$ of the reciprocal lattice. This function satisfies the periodicity condition

$$U(\mathbf{r} + \boldsymbol{\rho}) = U(\mathbf{r}) \tag{2.8}$$

Here $\boldsymbol{\rho}$ is a lattice vector in the form of (2.1). We note that for the lattice vectors $\boldsymbol{\rho}$ the reciprocal-lattice vectors $\mathbf{G}$ satisfy the relation

$$e^{i\mathbf{G}\boldsymbol{\rho}} = 1 \tag{2.9}$$

We see this by writing

$$\mathbf{G}\boldsymbol{\rho} = (h_1\mathbf{A} + h_2\mathbf{B} + h_3\mathbf{C})(n_1\mathbf{a} + n_2\mathbf{b} + n_3\mathbf{c}) = 2\pi(h_1 n_1 + h_2 n_2 + h_3 n_3).$$

This integer multiple of 2π directly yields the periodicity (2.8):

$$U(\mathbf{r}+\boldsymbol{\rho})\sum_{\mathbf{G}} u_G e^{i\mathbf{G}(\mathbf{r}+\boldsymbol{\rho})} = \sum_{\mathbf{G}} u_G e^{i\mathbf{G}\mathbf{r}} = U(\mathbf{r})$$

We turn now to the diffraction problem, scattering a wave from the incoming wave vector $\mathbf{k}$ into the outgoing wave vector $\mathbf{k}'$. The intensity of the scattered wave is proportional to

$$\text{intensity} \sim \int d\tau\, e^{i\mathbf{r}(\mathbf{k}-\mathbf{k}')}U(\mathbf{r}) = \sum_{\mathbf{G}} u_G \int d\tau\, e^{i\mathbf{r}(\mathbf{k}-\mathbf{k}')}e^{i\mathbf{G}\mathbf{r}}$$

($d\tau$ = volume element). The last integral vanishes due to cancellation by interference, except for the case when the exponent itself vanishes, i.e., when $\mathbf{k} - \mathbf{k}' + \mathbf{G} = 0$. So we find the important new form of the Bragg condition for the constructive interference between two waves which are scattered by the crystal from $\mathbf{k}$ to $\mathbf{k}'$:

$$\mathbf{k}' - \mathbf{k} = \mathbf{G} \qquad\qquad (2.10)$$

From the simple form of (2.7) and (2.10) we see again the advantage gained by the concept of the reciprocal lattice.

The abstract momentum space ($\mathbf{k}$-space) with the reciprocal lattice is divided into Brillouin zones. The boundaries of the Brillouin zones are found by erecting a plane in perpendicular direction at exactly the middle of a connecting straight line to a point of the reciprocal lattice. If one uses larger and larger reciprocal lattice vectors in this construction, one obtains the first, second, third, etc. Brillouin zone. In Fig. 2.9 we show as an example the first Brillouin zone in the two-dimensional

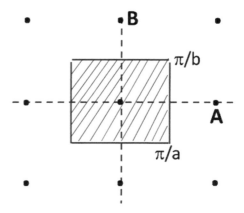

Fig. 2.9 Construction of the first Brillouin-zone in the (two-dimensional) case of the two fundamental vectors **A** and **B** of the reciprocal lattice oriented perpendicular to each other

case of the two fundamental vectors **A** and **B** oriented perpendicular to each other. As we will see in Chap. 4, in the case of the electronic band structure of materials the Brillouin zones play an important role.

Following the initial success of X-ray diffraction experiments, the method was quickly developed further along different directions. In the "rotating-crystal technique" a well-focused monochromatic X-ray beam is directed upon the crystal and, simultaneously, the crystal is rotated around a fixed axis. The high intensity of the diffracted radiation is observed only for distinct angle orientations of the crystal relative to the incoming X-ray beam, for which the Bragg interference condition is satisfied.

The two X-ray diffraction techniques that we have discussed so far require sufficiently large single crystals. As proposed for the first time by Peter Debye and Paul Scherrer, even crystal powder can be used. The powder may be compressed into the form of a little cylinder. For this powder technique one again uses monochromatic X-rays. Among the many randomly arranged little crystals in the powder there always exist a sufficiently large number for which the Bragg diffraction condition is well satisfied by their orientation. Here the sample rotation of the rotating-crystal technique is done quasi-automatically. As an important result we note that all three methods serve well for exactly determining the atomic or molecular distances between nearest neighbors in the crystal lattice, if the wave length of the X-rays is known.

During his experiments in Würzburg, Röntgen discovered the new radiation when he was investigating the physical behavior of gas discharges within an evacuated cathode ray tube. After only a few weeks of intensive experimentation he found that the new radiation always appeared if the fast electrons were abruptly decelerated by a solid obstacle in the glass tube. Here objects made from heavy elements such as tungsten or platinum were particularly effective. The principle of the generation of "bremsstrahlung", as it was subsequently called, had been found, and it continues to be used today in the construction of X-ray sources. Röntgen had received one of his first glass tubes, especially designed for the generation of X-rays with fused cathode and anode, from the glassworks of the Company "Greiner and Friedrichs" in the small town of Stützerbach near Ilmeau in Thuringia. Eventually, the large companies of the electronics industry took up the manufacture of X-ray equipment, and this field developed into an important business sector. Right up until today, the degree of automatization and standard of operation of the equipment has continuously improved.

The generation of X-rays within large rings-shaped electron accelerators, referred to as electron synchrotrons, is the latest development. An impressive example is the large European Synchrotron Radiation Facility (ESRF) in Grenoble with its ring diameter of 270 m (Fig. 2.10). Along the circular structure there is room for about 60 different measuring stations (beam-ports). The accelerated electrons move along a circular trajectory at a high energy of 6 GeV. This trajectory results from the balance between a force directed towards the center of the ring and the centrifugal force directed outwards. Because of this constant acceleration of the electrons in order to keep their circular trajectory, "synchrotron radiation" is emitted, the frequency of which depends upon the kinetic energy of the electrons. By means of special deflection elements inserted into the ring, the so-called

Fig. 2.10 Photograph of the large European synchrotron radiation source in Grenoble (bright ring in the foreground). The round tower-like container next to the ring is the external envelope of the German-French Research Reactor of the Laue-Langevin Institute. (Studio de la Révirée, Grenoble)

wigglers or undulators, individual beams with special properties can be supplied to the different beam-ports. Today, large intensive radiation sources similar to that in Grenoble are in operation worldwide at several locations. These radiation sources mainly serve to generate electromagnetic radiation of high intensity in the far ultraviolet and in the near X-ray spectral range.

Today, X-ray diffraction represents one of the most important tools for the analysis of materials. As outstanding indications of the importance of X-rays in determining the structure of materials, in addition to the first experimental proof of the lattice structure of crystals discussed before, we mention, for example, the X-ray analysis of Max Perutz of the hemoglobin in red blood cells which provides human oxygen transport, and the famous proposal for the structure of the DNA double helix by Francis Harry Compton Crick and James D. Watson, which during winter 1952/1953 was confirmed for the first time by Rosalind Franklin with her X-ray images. At this point we must also mention the work by Robert Huber, Johann Deisenhofer, and Hartmut Michel, clarifying the three-dimensional structure of the reaction center of photosynthesis by means of X-ray diffraction. The last three examples emphasize the great importance of X-rays in analyzing the structure of complex organic materials such as proteins and nucleic acids. Therefore, these concepts dating back to Max von Laue, W. H. Bragg, and W. L. Bragg, have been increasingly refined. In this way it has become possible that, from the patterns of

X-ray diffraction, both the periodic spatial lattice arrangement of the molecular crystal and the inner atomic structural detail of the protein molecules, can be reconstructed. However, in this case the relevant organic crystals must be prepared, which can often prove difficult and can require special attention.

In addition to the diffraction of X-rays, we again mention the (elastic) scattering of neutrons and its increasing importance for the analysis of the structure of crystals. In recent years in several research centers, the facilities for neutron diffraction have been strongly expanded.

Finally, we wish to discuss a phenomenon, the discoverer of which at first experienced rejection and mockery, until, after the gradual acceptance of his ideas, in 2011 he became honored by the award of the Nobel Prize in Chemistry: Daniel Shechtman and his discovery of quasi-crystals. On April 8. 1982 at the Johns Hopkins University in Baltimore, USA, with his electron microscope he obtained a diffraction image of an alloy of aluminum and manganese showing a tenfold symmetry. Hence, during a complete rotation an identical image was reproduced ten times, i.e., after each additional 36 degrees. The further research indicated, that the crystal itself exhibited a fivefold symmetry, which should be impossible according to the state of knowledge at the time. In such a crystal showing fivefold symmetry the atoms are arranged regularly, but not exactly periodically any more. How could Shechtman obtain the sharp diffraction pattern?

For his results Shechtman was strongly criticized from all directions. Even he had to leave his research group in Baltimore, and went back to Israel. He could publish his results only two and a half years later. However, subsequently, support for Shechtman started gradually. Five weeks after his paper there appeared a publication discussing this new type of crystals, and the name "quasi-crystals" was proposed. Already in the middle 1970s the mathematician Roger Penrose had demonstrated, that, by using only two building blocks, a slender and a compressed, skewed rectangle, one can generate a pattern completely covering a plain surface, but never exactly repeating itself. In the meantime, quasi-crystals have been found in many materials systems and also in nature.

In its communication regarding Shechtman, the Nobel Committee emphasized especially the "important lesson for science":

> Apparently, the achievement by Dan Shechtman not only consists of the discovery of the quasi-crystals, but also of the fact, that he recognized the importance of this discovery and resolutely conveyed it to a skeptical scientific community.

2.3 Types of Bonding

In the last part of this chapter we want to discuss the types of bonding between the (atomic or molecular) constituents of a solid. In principle, in this case we always deal with some kind of an attractive (electrostatic) interaction between electrons and the positively charged atomic nuclei (gravity and nuclear forces do not play any role in this discussion). There are special cases possible, resulting in the different types of bonding. In the form of a solid (or of condensed matter) the total energy of all constituents is always lower than in the case when it is completely disassembled into its atoms or molecules without any mutual interactions any more. The corresponding energy gain represents the binding energy.

One distinguishes between the following main types of bonding: van-der-Waals bond, ionic bond, covalent bond, metallic bond, and hydrogen bond.

Starting with the **van-der-Waals bond**, we consider atoms with completed electron shells (for example noble gases), which are difficult to ionize. On time average, at each atom we find a spherically symmetric charge distribution of the negative electron shells around the positive atomic nucleus. Whereas in the stationary state the electric field in the external space vanishes, the motion of the outer electrons (fluctuations) results in a temporally varying dipole moment (the time-averaged value of which remains zero). Considering two atoms (1) and (2) at distance R, the instantaneous dipole moment p_1 of (1) generates the electric field

$$|\mathbf{E}| = 2p_1/R^3 \qquad (2.11)$$

at the location of (2). The result is the induced dipole moment

$$p_2 = \alpha|\mathbf{E}| = 2\alpha\,p_1/R^3 \qquad (2.12)$$

of (2). Here α denotes the polarizability. The attractive interaction between $\mathbf{p}_1$ and $\mathbf{p}_2$ can be seen from the two cases shown in Fig. 2.11. In case 1, because of the decrease of the electric field $\mathbf{E}$ with increasing distance of $\mathbf{p}_1$, there results a force on $\mathbf{p}_2$ directed to the left. In case 2 we have an attraction again, since the distances between the opposite charges (attracting each other) are smaller than those between the same charges (repelling each other).

Fig. 2.11 Attractive interaction between the dipoles $\mathbf{p}_1$ and $\mathbf{p}_2$ of the two atoms (1) and (2) in the case of the van-der-Waals bond. Further details in the text

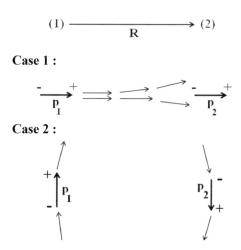

As we know from electrostatics, the potential energy U(**R**) between two dipoles **p**₁ and **p**₂ at mutual distance **R** is

$$U(\mathbf{R}) = \frac{\mathbf{p}_1 \mathbf{p}_2}{\mathbf{R}^3} - \frac{3(\mathbf{p}_1 \mathbf{R})(\mathbf{p}_2 \mathbf{R})}{\mathbf{R}^5}. \tag{2.13}$$

In case 1 of Fig. 2.11 one obtains

$$U(\mathbf{R}) = -2\mathbf{p}_1 \mathbf{p}_2 / \mathbf{R}^3 = -4\alpha\, \mathbf{p}_1^2 / \mathbf{R}^6. \tag{2.14}$$

Here we note, that the temporal average $\overline{\mathbf{p}_1^2} > 0$, even though $\overline{\mathbf{p}_1} = 0$.

The weak van-der-Waals bond is undirected. It is found in noble gases and in crystals of many organic molecules. The value of the bond energy is about $U \approx 10^{-21}$ J $\approx 10^{-2}$ eV with R $\approx$ 0.4 nm.

At this point we include a remark about energy units:

We have: 1 eV (electron volt) = $1.6 \cdot 10^{-19}$ J (Joule) and - with Boltzmann's constant k_B - the thermal energy $k_B T \approx 10^{-23} \frac{J}{K} T$. Hence, the value of U given above is obtained in the case T = 100 K. Therefore, the melting temperature of these materials is about 100 K.

From (2.13) and (2.14) one finds that the bond energy U(R) decreases rapidly with increasing distance R, proportional to R^{-6}. However, because of the overlapping of the electron shells, at sufficiently small distance one finds repulsion. Assuming the empirical repulsion law proportional to R^{-12} one obtains

$$U(R) = 4\varepsilon \left[\left(\frac{\sigma}{R} \right)^{12} - \left(\frac{\sigma}{R} \right)^6 \right]. \tag{2.15}$$

The model potential (2.15) fixed by means of the two parameters ε and σ is referred to as the Lennard-Jones potential. It is shown in Fig. 2.12. This potential and other similar model potentials having two or more parameters adjusted to the experimental data serve for calculating physical quantities.

Fig. 2.12 Schematic of the Lennard-Jones potential U(R) according to (2.15)

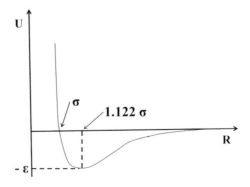

Turning now to a crystal consisting of N atoms, with (2.15) one finds the total energy U_{tot} by summation over all pairs:

$$U_{tot} = \frac{1}{2} N \, 4\varepsilon \left\{ \sum_j \left(\frac{\sigma}{R \, \rho_{ij}} \right)^{12} - \sum_j \left(\frac{\sigma}{R \, \rho_{ij}} \right)^6 \right\}. \qquad (2.16)$$

The i-th atom is arbitrarily selected, and the sum $\sum_j (-)$ runs over all $j \neq i$. For all possible values of i one obtains the factor N. Since in this way every atom is counted   twice,   the   factor   ½   must   be   applied.   (The   numbers $\rho_{ij} = 1, 2, 3, \ldots, \sqrt{2}, \sqrt{3}, \ldots$, etc. are introduced in order to make the final result independent of the special substance).

In the case of a cubic face-centered crystal one obtains

$$\sum_j \left(\frac{1}{\rho_{ij}} \right)^{12} = 12.131, \quad \sum_j \left(\frac{1}{\rho_{ij}} \right)^6 = 14.454. \qquad (2.17)$$

In the cubic face-centered lattice there are twelve nearest neighbors. This fact together with (2.17) indicates to us the rapid convergence of the sums. From the condition $\partial U_{tot}/\partial R = 0$ one obtains the equilibrium distance $R_o = 1.09 \, \sigma$, in good agreement   with   the   values   of   the   noble   gases.   Furthermore,   one   finds $U_{tot}(R_o) = -2.15 \, (4 \, N\varepsilon) = -8.6 \, N \, \varepsilon$. Based only on the twelve nearest neighbors, we expect $U_{tot} = -6 \, N \, \varepsilon$.

Next we turn to the **ionic bond**. The ionic crystals are composed of positive and negative ions with completed electron shells. In the crystal lattice the charges of opposite sign are alternately arranged next to each other, such that the attractive Coulomb force between opposite charges predominates the repulsion between equal charges. As an example we mention table salt (NaCl) with the positive Na^+- and the negative Cl^-- ions. In the case of the distance $R_o \approx 0.3$ nm between the charges, the electrostatic interaction energy amounts to $e^2/R_o \approx 5$ eV (e = elementary charge). As we can see, the bond energy is distinctly higher than in the case of the van-der-Waals bond.

The R^{-1}-dependence of the bond energy results in a long range of the interaction and, hence, in an interaction between charges of the opposite as well as of the same sign. Again, in the limit of small distances one expects repulsion. The repulsion law can be chosen in the form $\lambda \, e^{-r_{ij}/\rho}$, where r_{ij} is the distance between ions i and j, and where ρ determines the drop of the potential.

Again, we find the total energy of a crystal consisting of N ions by selecting an ion i and by summing over all $j \neq i$:

$$U_{tot} = \frac{N}{2} \sum_{j} \left(\lambda e^{-r_{ij}/\rho} \mp \frac{e^2}{r_{ij}} \right). \tag{2.18}$$

The first part in the bracket contains the repulsion and the second part the Coulomb interaction between the ions. With R denoting the distance between neighbors, and writing $r_{ij} = R\rho_{ij}$, from (2.18) we obtain:

$$U_{tot} = \frac{N}{2} z \lambda e^{-R/\rho} - \frac{N e^2}{2 R} \left(\sum_{j} \frac{(\mp 1)}{\rho_{ij}} \right). \tag{2.19}$$

In (2.19) in the contribution of the repulsion we have included only the number z of the nearest neighbors. The sum in the bracket

$$\left(\sum_{j} \frac{(\mp 1)}{\rho_{ij}} \right) = \alpha \tag{2.20}$$

is referred to as the Madelung constant, which depends on the crystal structure. Hence, we obtain:

$$U_{tot} = \frac{N}{2} \left(z \lambda e^{-R/\rho} - \alpha \frac{e^2}{R} \right). \tag{2.21}$$

In the case of table salt one finds $\alpha \approx 1.7$. In the calculation of α, sometimes the convergence is slow, and summation tricks can be helpful.

Again, the equilibrium distance R_o is found from the condition $\partial U_{tot}/\partial R = 0$, finally yielding the energy

$$U_{tot} = -\frac{N \alpha e^2}{2 R_o} \left(1 - \frac{\rho}{R_o} \right). \tag{2.22}$$

One finds $\rho/R_o \approx 0.1$, which indicates the small range of the repulsive force. The expression in front of the bracket is referred to as Madelung energy.

Erwin Rudolf Madelung has been the first publishing such calculations in 1918. From 1921 until his retirement in 1949 he was the Director of the Institute of Theoretical Physics at the University of Frankfurt.

The **covalent bond** represents the main type of bond in chemistry and in particular in organic chemistry. A very strong covalent bond is found between two carbon atoms in diamond with its bond energy of 7.3 eV. The pronounced directional dependence is characteristic, for example, in the case of the tetrahedral diamond structure of carbon, silicon, and germanium.

The covalent bond between two atoms occurs by means of a common pair of electrons having opposite spin. A simple example is the hydrogen molecule (H_2).

The theoretical understanding had to wait for quantum mechanics created in the 1920s. In their famous paper from 1927, Walter Heinrich Heitler and Fritz London have applied quantum mechanics in the case of the covalent bond of the hydrogen molecule. Here the central point is the symmetry property of the wave function regarding the exchange of both particles and their exchange interaction, resulting from the indistinguishability (identity) of both electrons. Their paper is fundamental for the valence-structure theory of quantum chemistry. Subsequently, in particular Linus Carl Pauling has strongly shaped the field of quantum chemistry.

The **metallic bond** can be understood in terms of a continuing development of the covalent bond, where a "free bond electron" is distributed over many atoms. The high electric conductivity of metals represents an important indication of the relatively free motion of the conduction electrons (one to two electrons per atom) within the lattice of the positive ions. In the quantum-mechanical theory the different contributions to the energy are taken into account:

$$\sum_i \frac{p_i^2}{2m} + \sum_i \sum_j \frac{e^2}{|R_i - R_j|} + \sum_i \sum_j \frac{e^2}{|r_i - r_j|} + \sum_{ij} V\big(|R_i - r_j|\big). \qquad (2.23)$$

From left to right the following contributions are listed in (2.23): (1) the kinetic energy of all conduction electrons, (2) the electrostatic interaction between the ions, (3) the electrostatic electron-electron interaction, and (4) the electron energy in the potential V of the ions. Here the theory has to deal with a "many-body problem", where great advances have been achieved during the last decades.

The **hydrogen bond** plays a role in certain chemical environments of hydrogen atoms, where the latter have transferred their electron to strongly electronegative neighboring atoms, such as, for example, fluorine, oxygen, or nitrogen. In this way an ionic-like bond appears. Because of the small diameter of a proton, this bond is only possible between two neighbors. It appears between the H_2O-molecules in the case of ice crystals, as well as in a large number of organic compounds. The bond between the two chains of the DNA double-helix represents a prominent example.

> **Regarding the increasing strength of the types of bonding we note the following sequence:**
>
> **van-der-Waals bond < hydrogen bond < metallic bond < ionic bond < covalent bond.**

Chapter 3
Permanent Movement in the Crystal Lattice

Abstract The quantized energy of the vibrations of the atoms in the crystal lattice contributes to the specific heat and the thermal conductivity. As noted first by Albert Einstein, the energy spectrum of the phonons is given by Planck's radiation law, resulting in a distinct deviation from the classic law of Dulong and Petit. Subsequently, the Einstein model, based on a single phonon frequency, was extended by Peter Debye by including the complete phonon spectrum.

On close inspection, the structure of a crystal does not represent a mathematically ideal point lattice. Instead, the atomic or molecular building blocks are permanently in motion. In some sense the crystal behaves more like a humming swarm of bees, where all bees still occupy spatially well ordered lattice sites. In a crystal each atom or molecule oscillates around a temporally average value of its spatial coordinates. In a popular model one imagines the crystal in the form of a three-dimensional lattice of point-like masses, where two neighboring points are connected with each other by little spiral springs. The total vibrational behavior of this three-dimensional arrangement of point-like masses and spiral springs can be separated into the complete set of elementary oscillations, referred to as "the normal modes", which are very useful in describing the dynamic state of the crystal. Each individual normal mode represents a "degree of freedom" of the crystal. At a temperature T, each degree of freedom carries the energy $k_B T$, where k_B denotes Boltzmann's constant. A crystal consisting of N atoms has 3 N vibrational degrees of freedom. Hence, the total vibrational energy U of the crystal amounts to $U = 3 N k_B T$. This relation is also referred to as the law of Dulong and Petit. The prefactor $3 N k_B$ indicates the heat capacity arising from the lattice vibrations, which is independent of temperature, according to this law. Already in 1819 the two Frenchmen Pierre Louis Dulong and Alexis Thérèse Petit had noted, that the specific heat of solids is nearly independent of temperature and connected with their molar mass.

© Springer Nature Switzerland AG 2019
R. P. Huebener, *Conductors, Semiconductors,*
Superconductors, Undergraduate Lecture Notes in Physics,
https://doi.org/10.1007/978-3-030-31420-0_3

3.1   Quantum Theory: Max Planck and Albert Einstein

So far we have restricted the discussion to the classical limit, and we have ignored quantum theory. Next we will address quantum theory. As was shown for the first time by Max Planck, the energy of light and heat radiation is quantized, each energy quantum having the energy $E = h\nu$, which is proportional to the frequency ν of the radiation. Here we have introduced Planck's constant h, which is a fundamental constant in physics. In his theoretical considerations, achieving a first result during December 1900, Planck just took the consequent conclusions from very precise optical measurements published shortly before.

As a final result of these efforts Planck was able to formulate his famous radiation law. In this way he succeeded in unifying the two theoretical laws which were already known but valid only in certain limiting cases: namely Wien's radiation law in the limit of small wavelengths and the Rayleigh-Jeans radiation law in the limit of large wavelengths (Fig. 3.1).

> The optical measurements had been carried out by a group of physicists (Ludwig Holborn, Otto Lummer, Ernst Pringsheim, Heinrich Rubens, Wilhelm Wien, and others) at the *Physikalisch-Technische Reichsanstalt* (German Bureau of Standards) in Berlin Charlottenburg (Fig. 3.2). At that time, in the *Reichsanstalt*, which was founded in 1887 mainly at the initiative of Werner Siemens and Hermann von Helmholtz, a reliable standard for the light intensity of the radiation emitted by hot and glowing pieces of metal was intended to be developed in a basic research program. This subject had become important because of the rapid spreading of the artificial lighting technology.

It was Albert Einstein, who in 1905 for the first time strictly applied the concept of the energy quantum to the propagation of electromagnetic waves and who introduced the idea of light quanta or photons, as they are called also (Fig. 3.3). Based on these concepts he could convincingly explain the photoelectric effect. Due to this effect, during irradiation with light of a frequency above a certain limiting value, electrons are emitted from the surface of metals. The energy of the emitted electrons only depends on the light frequency, whereas the light intensity only

Fig. 3.1 The radiation laws of the black body according to Planck, Wien, as well as Rayleigh and Jeans. Spectral radiation intensity plotted versus the frequency

Fig. 3.2 Radiation test laboratory in the *Physikalisch-Technische Reichsanstalt* in Berlin around 1900. (*Photo* Physikalisch-Technische Bundesanstalt Braunschweig and Berlin)

affects the number of the emitted electrons. In 1921 Einstein was awarded the Nobel Prize in Physics for his theory of the photoelectric effect. Subsequently, Einstein's hypothesis of the light quanta has been confirmed impressively by many experiments.

How strongly Einstein's ideas have revolutionized the physical thinking at the beginning of the last century, is well illustrated by the following quotation from the document written by Max Planck on June 12, 1913, proposing Albert Einstein to become an Ordinary Member of the Prussian Academy of Science:

... In summary one can say, that among the great problems, of which modern physics is so rich, there is hardly one, about which Einstein had not expressed an opinion in an important way. That in his speculations occasionally he may have gone too far, as for example in his hypothesis of the light quanta, one must not blame on him too heavily; since without daring to take a risk once in a while, a true advancement cannot be introduced also in the exact sciences. ...

This document carries the signatures of Planck, Nernst, Rubens, and Emil Warburg.

Fig. 3.3 Max Planck (left) and Albert Einstein (right). (*Photos* Deutsches Museum, München)

3.2 Specific Heat of the Crystal Lattice, Phonon Spectrum

Following these remarks on quantum theory, we return to the crystal lattice. The energy quantization of the electromagnetic waves should apply similarly also to vibrations in the crystal lattice. Again, it was Albert Einstein who took up this idea for the first time in 1906. He proposed that the elements at each site of the crystal lattice oscillate with a single frequency, the Einstein frequency $v_E = \omega_E/2\pi$, and that the vibrational energy is quantized again in units $E = h v_E = \hbar \omega_E$ (with $\hbar = h/2\pi$). The quanta of the vibrational energy in crystals are referred to as phonons. According to the Einstein model, the crystal energy U due to the lattice vibrations is given by

$$U = 3\,N < n_\omega > \hbar \omega \tag{3.1}$$

Here we assume, that the crystal consists of N atoms. Hence, there are 3N vibrational degrees of freedom. $< n_\omega >$ is the probability, that a vibrational state with the angular frequency ω and the quantized energie $\hbar \omega$ is occupied. This probability is given by the Bose-Einstein distribution

$$< n_\omega > \; = \frac{1}{e^{\hbar \omega / k_B T} - 1} \tag{3.2}$$

to which we will return below again. The distribution (3.2) had been proposed by Planck for the first time in his famous law of electromagnetic radiation.

In his model of the quantized energy of the lattice vibrations, in (3.1) and (3.2) Einstein at first had assumed only a single frequency $\omega = \omega_E$. Only a few years later, this Einstein model has been extended by Peter Debye, who assumed a continuous frequency spectrum of the vibrations, ranging between zero and a characteristic maximum frequency, the Debye-frequency ω_D. Now the vibrational energy of the crystal is given as an integral over all phonon frequencies

$$U = \int_0^{\omega_D} d\omega \, D(\omega) <n_\omega> \hbar\omega \tag{3.3}$$

Here $D(\omega)$ is the number of vibrations per frequency interval, also referred to as density of states. (In Chap. 6 we will return to the calculation of the density of states $D(\omega)$). As an example, in Fig. 3.4 we show the spectral energy density of the phonons of a germanium crystal.

In this way Debye was able for the first time to explain the temperature dependence of the total energy of the lattice vibrations in crystals and in particular the famous T^3 behavior of the specific heat at low temperatures, in excellent agreement with experiment. Again, Planck's energy quantization, now applied to

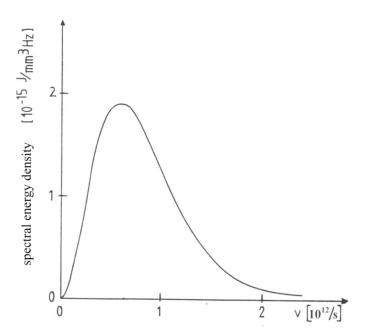

Fig. 3.4 Spectral energy density of the phonons in a germanium crystal at the temperature of 10 K according to the Planck radiation law, plotted versus the phonon frequency ν

the lattice vibrations, has played a central role, and the classical law of Dulong and Petit has been eliminated.

The measurements of the specific heat of crystals, together with the Debye model, had created a significant advance of the general acceptance of quantum theory. For example, it was only after these measurements that the later Nobel Laureate Walther Nernst became convinced that Planck's quantum theory was more than just an interpolation formula and that it represented new fundamental physics. The fact, that Planck had based his revolutionary new idea on the physics of heat radiation, had resulted from the relatively high level of experimental optics already reached at that time.

> In the years 1910–1916 in Berlin, in an extensive research program together with many collaborators, Walther Nernst systematically investigated the specific heat of many solid materials at low temperatures, confirming the quantum theory. In the meantime, Nernst had recognized clearly the fundamental importance of quantum theory. Therefore, he also organized the first large conference dealing with this subject, the famous First Solvay-Conference on *"The Theory of radiation and the quanta"* held in Brussels from October 30 until November 3, 1911. For this conference Nernst had been able to enlist the financial support of the Belgian industrialist Ernest Solvay.

The important central idea during the development of quantum mechanics in the 1920s was the strict limitation of all statements about the atomic world to observable facts. The fact that the elementary particles such as electrons, protons, neutrons, etc., are exactly identical, must be centrally incorporated into the theory. If the same two elementary particles are exchanged, the result must remain unaffected. Hence, the theoretical qualities must have certain symmetry properties. This requirement has severe consequences for the probability distribution of the different states of the systems, and new concepts for quantum statistics are needed. The first steps in this direction originated from the Indian physicist Satyendra Nath Bose. In 1924 he had derived Planck's radiation law in a new way. Since he ran into difficulties during the publication of his results, he approached Einstein asking him for support. Einstein felt enthusiastic about Bose's paper and arranged for its publication in the *Zeitschrift für Physik*. Subsequently, starting from Bose's results, in some additional papers Einstein pointed out the formal similarity between radiation and an ideal gas. Today, the resulting concept of quantum statistics is referred to as Bose-Einstein statistics.

Bose-Einstein statistics apply to exactly identical elementary particles with zero or integer angular momentum. Hence, these particles are also called bosons. A single quantum state can be occupied by an arbitrarily large number of bosons. Since phonons have zero angular momentum, they belong to this kind of particles. Light quanta or photons are bosons also, since their angular momentum is equal to one. However, electrons require a different kind of quantum statistics, as we will discuss in Chap. 5.

The energy spectrum of phonons is described also by Planck's radiation law, similar to the photon spectrum of a heat radiator (Fig. 3.4). However, compared with the spectrum of electromagnetic radiation, the phonon spectrum displays an important difference, since its frequencies are restricted to the range below a characteristic maximum frequency, the "Debye frequency" ω_D. This maximum frequency ω_D simply results from the discrete lattice structure of the crystals. Below the nearest-neighbor distance of the crystal lattice, length scales for the lattice vibrations are meaningless, resulting in the definition of a minimum value of the wavelength and a corresponding maximum value of the vibration frequency. The resulting maximum value $\hbar\,\omega_D$ of the phonon energy is referred to as the Debye energy. Since, on the other hand, the electromagnetic waves propagate in a continuous medium without any lattice structure, in this case we are not confronted with corresponding minimum or maximum values of the wavelength and the frequency, respectively.

One finds the density of states $D(\omega)$ from the number of the elastic natural frequencies fitting exactly into the volume of the crystal, say, into a cube with the edge length L. Furthermore, in K-space the increment of the volume due to the frequency increment $d\omega$ must be considered. (Here and in the following, **K** denotes the phonon wave vector). In this way one finds $D(\omega) \sim \omega^2$. By inserting $< n_\omega >$ from (3.2) into (3.3), finally we obtain the vibrational energy of the crystal. The specific heat (at constant volume) of the lattice vibrations derived from this energy is given by

$$C_V = (\frac{\partial U}{\partial T})_V = 9\,N\,k_B \left(\frac{T}{\theta}\right)^3 \int_0^{z_D} dz\, \frac{z^4 e^z}{(e^z - 1)^2} \qquad (3.4)$$

Here, $\theta = \hbar\omega_D/k_B$ denotes the Debye-temperature. In this context, introducing the quantity $z = \hbar\omega/k_B T$ is convenient ($z_D = \hbar\omega_D/k_B T = \theta/T$). At low temperatures ($T << \theta$) the expression in (3.4) yields $C_V \sim T^3$, in good agreement with the experimental data. On the other hand, at high temperatures ($T >> \theta$) from (3.4) one obtains the constant result $C_V = $ const, i.e., the classic law of Dulong and Petit.

The dependence of the phonon frequency ω on the wave vector **K** of the phonons is obtained from a theoretical model, in which the crystal lattice is approximated by point-like masses connected with each other by elastic spiral springs. In the simplest case of a linear chain with the lattice constant a and the same point-like masses m, connected with each other by means of the same spring constant f, one finds the dispersion relation

$$\omega = \left(\frac{4f}{m}\right)^{1/2} \sin\frac{K\,a}{2} \qquad (3.5)$$

In the case of a discrete lattice with the distance a between neighbors, the wave vector **K** is confined to the range of the first Brillouin zone, i.e., in the case of the one-dimensional chain to the range $-\frac{\pi}{a} \le K \le \frac{\pi}{a}$. In the limit of small values of the wave vector, one obtains from (3.5):

$$\omega = \left(\frac{a^2 f}{m}\right)^{1/2} K \qquad (3.6)$$

The factor $\left(\frac{a^2 f}{m}\right)^{1/2}$ represents the sound velocity. Equation (3.6) indicates the acoustic limit, and (3.5) characterizes an acoustic mode.

In the simplest case we only have to deal with a single atom per elementary cell of the crystal lattice. The vibration of the atom can occur within all three spatial dimensions. Hence, the phonons propagating through the crystal are also characterized by different spatial directions of the lattice vibrations. In the case of "longitudinal phonons" the atoms of the crystal lattice vibrate parallel to the propagation direction of the wave. On the other hand, for the "transverse phonons" the vibration occurs along each of the two principal directions perpendicular to the propagation direction, respectively. In this way the three "acoustic phonon modes" must be distinguished: one longitudinal mode and two transverse modes. They are referred to as acoustic modes, since in the limit of large wavelengths the corresponding phonons propagate with the velocity of sound (Fig. 3.5). (Because of the cubic face-centered symmetry in the case of the example of copper, the boundary of the first Brillouin zone is located at $K = 2\pi/a$, in contrast to $K = \pi/a$ in the case of the linear chain or the simple cubic symmetry with the lattice constant a).

If the crystal lattice contains more than a single atom per elementary cell, the "optical modes" must be added. In the case of the optical modes, the atoms within the elementary cell oscillate in opposite phase relative to each other. If the atoms carry opposite electric charge, as in the case of the ionic crystals, due to these oscillations electric dipole moments appear, which affect the optical properties. In the case of p atoms per elementary cell, there are a total of 3p different phonon modes: 3 acoustic modes and 3p − 3 optical modes.

Fig. 3.5 Phonon angular frequency $\omega = 2\pi v$ plotted as a function of the wave number K of the phonons ("dispersion curves") of a copper crystal. The wave vector is oriented along the direction of the cube edges of the cubic unit cell of the crystal. L denotes the longitudinal and T the transverse phonon branch. a = 0.361 nm is the distance between neighbors in the cubic crystal lattice of copper. (B. N. Brockhouse)

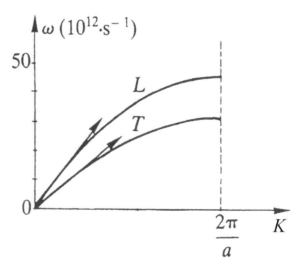

Experimentally, the energy spectra of the phonons are determined mostly by means of inelastic neutron scattering in crystals. The pioneering experiments were performed by Bertram Neville Brockhouse from McMaster University in Canada. He carried out his first measurements in 1955 with aluminum. One needs to measure the change in energy and momentum of the neutrons as they emit or absorb a phonon within the crystal. The impressive advances in the field of neutron spectroscopy, which to a large extent were due to Heinz Maier-Leibnitz and his group first at the Technical University of Munich in Garching and later at the Laue-Langevin Institute in Grenoble, have turned out to be extremely fruitful.

Max von Laue and his colleagues in Munich, prior to their famous X-ray diffraction experiment, were very concerned at the time by the following obvious question: Is it not likely that the perturbation of perfect order in the crystal lattice due to the permanent lattice vibrations really ruins the observation of X-ray diffraction? However, the clear and positive experimental result gave a decisive answer to this question. In 1913 Peter Debye wrote a series of papers theoretically treating the role of the thermally excited vibrations of the crystal lattice as it affects the diffraction of X-rays. During the 1920s this subject was taken up again by the Swedish theorist Ivar Waller. Both scientists have shown that the thermal lattice vibrations only effect a reduction of the maximum intensity of the diffracted beam, whereas the linewidth of the diffracted beam remains the same. The reason for this is simply that the thermal motion of the many atoms in the lattice is completely uncorrelated, leading effectively to a cancellation between the oscillations of the different atoms. This reduction of intensity of the diffracted beam is quantified in terms of the "Debye-Waller factor". This factor indicates, furthermore, that the intensity of the diffracted beam strongly increases with decreasing temperature.

The Debye-Waller factor played a key role later during the discovery of the Mössbauer effect. As a young Ph.D. student in Munich Rudolf L. Mössbauer has been asked by Heinz Maier-Leibnitz, his thesis advisor, to study the resonance absorption of γ-radiation in atomic nuclei. Mössbauer was supposed to find out if the γ-radiation emitted by an atomic nucleus of the source is resonantly re-absorbed by another atomic nucleus of the same element in the absorber. For the observation of this effect it is necessary for there to be sufficient overlap between the spectral energy widths of the γ-radiation for the emission and the absorption processes. Here the main issue centered around the question of whether this required overlap disappears, perhaps completely, because of the recoil during the emission and absorption of the γ-quantum, thereby eliminating the possibility of resonance absorption. It is interesting that initially in particular Heinz Maier-Leibnitz had the idea of enhancing this energetic overlap by increasing the temperature and thereby enlarging the thermal linewidth of the γ-radiation. However, this idea was soon discarded, and Mössbauer instead cooled the source and the absorber with liquid oxygen. This then turned out to be crucial, and Mössbauer could observe for the first time the recoil-free nuclear resonance fluorescence. He performed his first experiments using the 129-keV-radiation of the iridium isotope [191]Ir, with the iridium atoms implanted within a crystal. Again, it was the elimination of the influence of the phonons on the linewidth of the Mössbauer line, which had produced the effect he was looking for, exactly as prescribed by the Debye-Waller factor. The question of the recoil now became irrelevant, since the iridium atoms were solidly implanted within the host crystal. The extremely narrow energy width of the Mössbauer line is finally limited only because of the "natural line width", resulting from the finite lifetime of the quantum mechanical state due to the Heisenberg uncertainty relation. Because of this extremely narrow linewidth a large number of highly sensitive measurements has become possible in many areas. An early spectacular case is the detection of the energy change in the quanta of γ-radiation after they have travelled upwards a certain distance in height in the gravitational field of the Earth. In 1960 this was detected by the American physicists Robert V. Pound and Glen A. Rebka. In their experiment the distance in height travelled was 22.5 m.

3.3   Thermal Conductivity of the Crystal Lattice

In crystals phonons also contribute to the transport of heat energy. In electrical insulators they represent the only mechanism determining the heat conductivity. The phonon contribution κ_G to the heat conductivity of a crystal is closely connected with the lattice component C of the specific heat. This is indicated by the simple formula of the kinetic theory of κ_G:

$$\kappa_G = \frac{1}{3} v \, C \, \ell \tag{3.7}$$

Here v is the velocity and ℓ the mean free path of the phonons. As a model, we can think about phonons in terms of a gas experiencing many collisions between its particles. Because of the many collisions, in general the heat transport provided by the phonons is a diffusive process. With increasing temperature the number of thermally excited phonons, and hence the collision probability among the phonons, increases. As a result, with increasing temperature the heat conductivity decreases. However, at very low temperatures we have an exception. In this case, the number of phonons is very small, the collisions between them become unimportant, and it is only the number of phonons which matters. Now the mean free path ℓ is limited by the collisions with the surface of the crystal, and it becomes independent of the temperature. In this regime we have $C \sim T^3$ and because of (3.7) $\kappa_G \sim T^3$.

At higher temperatures, an exact theoretical treatment shows, that a novel process must be taken into account, the "umklapp-process" (U-process). It effects, that the directed momentum of the phonons is transferred to the crystal and is lost for the heat transport. Denoting the wave vectors of the phonons participating in the umklapp-process by $\mathbf{K}_i$, we obtain

$$\mathbf{K_1} + \mathbf{K_2} = \mathbf{K_3} \pm \mathbf{G}. \tag{3.8}$$

$\mathbf{G}$ is a vector in the reciprocal lattice. Processes with $\mathbf{G} = 0$ are referred to as N-processes. The concept of the umklapp-processes has been proposed for the first time in 1929 by Rudolf E. Peierls, who was born in Berlin and later emigrated to England. (In Chap. 2 in the case of (2.10) we have dealt already with a similar case, in which during a Bragg reflection the momentum of the photon is transferred to the crystal).

The umklapp-processes yield an important contribution only if $\mathbf{K_1} + \mathbf{K_2} \geq \frac{1}{2}\mathbf{G}$. However, at low temperatures they are frozen out, since phonons satisfying this condition are not thermally excited. On the other hand, at high temperatures ($T > \theta$) the U-processes are dominant. Their number increases proportional to the number of phonons, which in turn is proportional to temperature. Hence, for the mean free path ℓ of the phonons one finds $\ell \sim T^{-1}$. Because $C = $ const in this case, from (3.7) we obtain: $\kappa_G \sim T^{-1}$. The overall result for the temperature dependence of the heat conductivity of the lattice is a curve with a distinct maximum (Fig. 3.6). For example, in sapphire (Al_2O_3) this maximum is located near 30 K.

Diamond is a material with an extremely high heat conductivity. Therefore, a special effort has recently been concentrated on the development of thin diamond layers, which are highly interesting technologically for cooling purposes because of their large heat conductivity.

3.4   Ballistic Phonons

As we have discussed before, the collision processes between phonons become more and more rare at sufficiently low temperatures. In this regime phonons can propagate freely over distances as large as, say, about mm to cm with sound velocity. In this case we are dealing with ballistic phonons. The propagation of ballistic phonons can be observed easily, if a heat pulse is applied locally to the front surface of a well-cooled crystal. The generated phonon pulse can be detected locally at the back of the crystal after the proper time of flight (Fig. 3.7). For

Fig. 3.6 Temperature dependence of the heat conductivity of an electrical insulator (schematically)

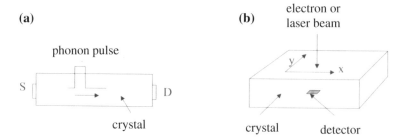

(a)

phonon pulse

(b) electron or laser beam

crystal   crystal   detector

Fig. 3.7 Ballistic propagation of phonons at low temperatures. **a** Scheme of an experiment to measure the propagation time of the phonons. On the left hand side of the crystal the phonons are generated by means of a heat pulse applied to the source S, and subsequently they are detected on the right hand side of the crystal with the detector D. A thin metal layer deposited on the crystal surface can act as a source by means of the application of an electric current pulse. A deposited thin layer, the electric resistance of which responds sensitively to pulsed temperature changes, can also serve as detector. **b** A pulsed electron beam or laser beam, directed at the crystal surface at one side of the crystal, can also be used for the generation of a phonon pulse. If furthermore the beam is scanned over the crystal surface, while the detector remains locally fixed on the opposite side of the crystal, the ballistic propagation of the phonons can be studied as a function of the propagation direction in the crystal

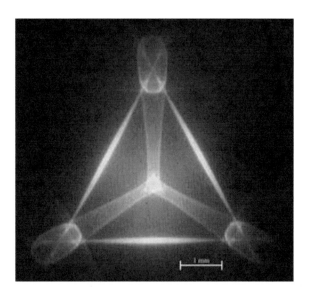

Fig. 3.8 Image of the intensity of the ballistic phonons in dependence on the propagation direction in a silicon single crystal at a temperature of 2.0 K ("phonon imaging"). While the crystal surface on one side is scanned with the electron beam, the intensity of the ballistic phonons is recorded on the opposite side of the crystal with a locally fixed detector in dependence on the coordinate point of the scanned crystal surface. Bright regions correspond to high intensity. The spatially diagonal line of the cubic unit cell of the crystal is oriented perpendicular to the scanned surface of the crystal

generation of the heat pulse, a pulsed laser beam or electron beam directed on the crystal surface can be used. Due to the anisotropy of the elastic crystal properties, the ballistic propagation of the phonon energy displays distinct maxima along certain directions within the crystal. This effect is referred to as phonon focusing, and can be easily demonstrated using the method we have just discussed. The only requirement is that the beam is scanned laterally over the crystal surface, while the detector is fixed locally at the back of the crystal. In Fig. 3.8 we show a typical example.

If the crystal is cooled to lower and lower temperatures, eventually only the "zero-point motion" of the building blocks of the crystal remains. The zero-point motion follows from the quantum mechanical uncertainty relation, which requires that a spatially fixed object always displays a finite uncertainty of its momentum. The resulting zero-point energy of an oscillator at frequency ν amounts to $\frac{1}{2} h \nu$.

Chapter 4
Electric Conductor or Insulator?— Energy Bands

Abstract Application of quantum mechanics to the electrons in the periodic potential of the crystal lattice yields the energy bands separated from each other by forbidden energy gaps. The approximations with bound electrons (F. Bloch) and with nearly-free electrons (R. Peierls) are discussed. Depending on how the energy states within the bands are occupied by electrons, we deal with electrical conductors, semiconductors, or insulators.

After the principles of the new quantum mechanics had been established during 1925 and 1926 mainly by Werner Heisenberg from Germany, Erwin Schrödinger from Austria, and Paul Adrien Maurice Dirac from England, there developed a strong interest in applying the theory to as many different cases as possible. Only by applying the theory to a large number of examples could familiarity with the new concepts be achieved. After simple cases such as the hydrogen atom or the hydrogen molecule had been treated, more complex problems were tackled. At that time important developments began in Leipzig.

In 1927, at the young age of only 26 years, Heisenberg had already accepted the offer of a Chair of Theoretical Physics at the University of Leipzig. Here he quickly attracted a group of extremely gifted and creative young scientists, who subsequently had a dominant impact on further developments in physics. At the beginning of 1928 Heisenberg had already recognized that quantum mechanics would play an important role for crystals. The Swiss scientist Felix Bloch, born in Zürich, had just joined Heisenberg's group as a Ph.D. student (Fig. 4.1). For his thesis Heisenberg proposed two possible subjects: Bloch could take up the quantum mechanical theory of ferromagnetism or the electron theory of metals. Since Bloch knew that Heisenberg had already worked out the basic parts of the first subject, he preferred the second one. Only in this way could he hope to come up with a significant contribution of his own. Indeed, soon afterwards Heisenberg published his famous theoretical paper which became the starting point of the modern theory of ferromagnetism.

The quantum mechanical theory of electrons in crystals requires the solution of the Schrödinger equation in the case of the spatially periodic crystal lattice. In this

© Springer Nature Switzerland AG 2019

R. P. Huebener, *Conductors, Semiconductors, Superconductors*, Undergraduate Lecture Notes in Physics, https://doi.org/10.1007/978-3-030-31420-0_4

Fig. 4.1 Werner Heisenberg (left), (Photo: Deutsches Museum) and Felix Bloch (right), (Photo: Nobel Museum)

case, the energy spectrum of the electrons is determined by the periodic potential U($\mathbf{r}$) of the crystal atoms. Then the Schrödinger equation is

$$-\frac{\hbar^2}{2m}\,\Delta\psi(\mathbf{r})\,+\,\mathrm{U}(\mathbf{r})\,\psi(\mathbf{r})=\varepsilon\,\psi(\mathbf{r}). \tag{4.1}$$

Here m denotes the mass and $\psi(\mathbf{r})$ the wave function of the electrons. Δ is the Laplace operator $\Delta\psi=\frac{\partial^2\psi}{\partial\mathrm{x}^2}+\frac{\partial^2\psi}{\partial\mathrm{y}^2}+\frac{\partial^2\psi}{\partial\mathrm{z}^2}$ and ε the energy of the electrons. The potential energy U($\mathbf{r}$) must satisfy the periodicity condition (2.8). In the theoretical treatment we distinguish between two different important approximations: the bound-state approximation (after Felix Bloch) and the free-electron approximation (after Rudolf E. Peierls). In both cases we deal only with the electrons in the highest available energy range, and not with the lower levels of the strongly bound states at the individual atoms at the sites of the crystal lattice.

4.1 Approximation with Bound Electrons (Felix Bloch)

In his dissertation, the results of which he published in 1928, Felix Bloch formulated the quantum mechanical foundation of the theory of electrons in crystal lattices. Due to the periodic potential of the crystal lattice in all three dimensions, the de Broglie waves of the electrons are spatially modulated following the rhythm of the lattice structure. In the approximation with bound electrons one assumes, that the electrons with the highest energy remain located most of the time at a certain lattice site, and only occasionally move to a neighboring lattice site, because of the small interaction between both locations. Their binding energy at a certain lattice site is assumed to be much larger than their kinetic energy. In this case the solution of the Schrödinger equation found by Bloch is based on the wave function $\varphi_O(\mathbf{r})$ of the quantum mechanical states of the electrons of the isolated individual atom or molecule with their corresponding discrete energy levels. We denote the coordinate of the electron by $\mathbf{r}$ and the coordinate of the atom or molecule (i.e., of the lattice site) by $\boldsymbol{\rho}$. The solution proposed by Bloch in form of the superposition of the atomic wave functions $\varphi_O(\mathbf{r}-\boldsymbol{\rho})$ is then given by:

$$\psi_k(\mathbf{r}) = \sum_{\rho} e^{ik\rho} \varphi_O(\mathbf{r}-\boldsymbol{\rho}). \tag{4.2}$$

This ansatz leads directly to the extended periodicity condition

$$\psi_k(\mathbf{r}+\boldsymbol{\rho}) = e^{ik\rho} \psi_k(\mathbf{r}) \tag{4.3}$$

with the phase factor $e^{ik\rho}$. The periodicity (4.3) can be seen by writing

$$\psi_k(\mathbf{r}+\boldsymbol{\rho_o}) = \sum_{\rho} e^{ik\rho} \varphi_O(\mathbf{r}+\boldsymbol{\rho_o}-\boldsymbol{\rho})$$

$$= e^{ik\rho_o} \sum_{\rho} e^{ik(\rho-\rho_0)} \varphi_O(\mathbf{r}-[\boldsymbol{\rho}-\boldsymbol{\rho_o}]) = e^{ik\rho_o} \psi_k(\mathbf{r})$$

The function $\psi_k(\mathbf{r})$ in (4.2) represents the famous Bloch ansatz for the quantum mechanical wave function of the electrons, upon which all further theoretical developments for crystals have since been built. The fact, that according to (4.3) a phase factor $e^{ik\rho}$ "...*can always be separated from the eigenfunctions, and where the rest only displays the periodicity of the lattice, can be vividly expressed such, that we are dealing with planar de-Broglie waves, which are modulated by the rhythm of the lattice structure*" (the words of Felix Bloch in his paper *About the Quantum Mechanics of the Electrons in Crystal Lattices*, published in 1928, here in English translation). With the wave function (4.2) one finds the electron energy ε_k in the following way. Applying the rules of quantum mechanics, we have

$$\int d\tau\, \psi_k^*(\mathbf{r})\varepsilon_k\psi_k(\mathbf{r}) = \varepsilon_k \int d\tau\, \psi_k^*(\mathbf{r})\psi_k(\mathbf{r}) \tag{4.4}$$

and using (4.1) we find

$$\varepsilon_k = \frac{1}{N}\int d\tau\, \psi_k^*(\mathbf{r})[-\frac{\hbar^2}{2m}\Delta + U_a(\mathbf{r}-\boldsymbol{\rho}) + U(\mathbf{r}) - U_a(\mathbf{r}-\boldsymbol{\rho})]\psi_k(\mathbf{r}) \tag{4.5}$$

Here $d\tau$ denotes the volume element. In (4.5) under the integral we have added and subtracted again the atomic potential $U_a(\mathbf{r})$ shown in Fig. 4.2, and we have used the normalization of the wave function

$$\int d\tau\, \psi_k^*(\mathbf{r})\psi_k(\mathbf{r}) = N \tag{4.6}$$

where N is the number of the crystal atoms.

We start with the first two terms under the integral in (4.5). We assume that the atomic functions $\varphi_O(\mathbf{r})$ of neighboring atoms overlap only very little. For each atomic function $\varphi_O(\mathbf{r} - \boldsymbol{\rho})$ in the Bloch sum the operator $[-\frac{\hbar^2}{2m}\Delta + U_a(\mathbf{r} - \boldsymbol{\rho})]$ yields the eigenvalue ε_o, and we obtain

$$\frac{1}{N}\int d\tau\, \varepsilon_o\psi_k^*(\mathbf{r})\psi_k(\mathbf{r}) = \varepsilon_o \tag{4.7}$$

The last two terms under the integral in (4.5) yield

$$\frac{1}{N}\int d\tau\, \psi_k^*(\mathbf{r})e^{i\mathbf{k}\boldsymbol{\rho}}[U(\mathbf{r}) - U_a(\mathbf{r}-\boldsymbol{\rho})]\varphi_O(\mathbf{r}-\boldsymbol{\rho}) =$$
$$= \frac{1}{N}\sum_{\boldsymbol{\rho}'}\left\{\sum_{\boldsymbol{\rho}}e^{i\mathbf{k}(\boldsymbol{\rho}-\boldsymbol{\rho}')}\int d\tau\varphi_o^*(\mathbf{r}-\boldsymbol{\rho}')[U(\mathbf{r}) - U_a(\mathbf{r} - \boldsymbol{\rho})]\varphi_o(\mathbf{r}-\boldsymbol{\rho})\right\}$$

In the double sum over $\boldsymbol{\rho}'$ and $\boldsymbol{\rho}$ the two indices are treated *independently* of each other, and we can use the following procedure. We keep one running index fixed and replace the sum by the factor N. For this we use $\boldsymbol{\rho} = 0$ (Since the summation index runs over all lattice points, an arbitrarily taken fixed index yields the same

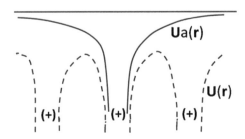

Fig. 4.2 Comparison of the energy of an electron, $U_a(\mathbf{r})$, in the case of an isolated individual atom (solid line), and the case of the presence of atoms at the neighboring lattice sites, $U(\mathbf{r})$ (dashed curve)

contribution as any other neighboring lattice point because of the translation symmetry). The factor N is cancelled, and we obtain for the last two terms (noting that we took $\boldsymbol{\rho} = 0$)

$$\sum_{\boldsymbol{\rho}'} e^{-ik\boldsymbol{\rho}'} \int d\tau \, \varphi_o^*(\mathbf{r}-\boldsymbol{\rho}')[U(\mathbf{r}) - U_a(\mathbf{r})]\varphi_O(\mathbf{r}) \tag{4.8}$$

In the sum over $\boldsymbol{\rho}'$ we start with $\boldsymbol{\rho}' = 0$, and then we take the nearest neighbors of $= 0$, i.e., $\boldsymbol{\rho}' = \pm a$, then the next-nearest neighbors $\boldsymbol{\rho}' = \pm 2a$, etc., where a denotes the lattice constant. However, often the next-nearest neighbors (and all others further away) can be neglected because of the vanishing product $\boldsymbol{\rho}_o^*(\mathbf{r}-\boldsymbol{\rho}') \, \varphi_O(\mathbf{r})$ indicating the overlap of the atomic functions.

The contribution $\boldsymbol{\rho}' = 0$ yields $\varepsilon_k = \varepsilon_o - \alpha$, with

$$-\alpha = \int d\tau \, \varphi_o^*(\mathbf{r})[U(\mathbf{r}) - U_a(\mathbf{r})]\varphi_o(\mathbf{r}) \tag{4.9}$$

$U_a(\mathbf{r})$ is the potential energy in the unperturbed individual atom. As we see from Fig. (4.2), due to the presence of the atoms at the neighboring lattice sites, we have $U_a(\mathbf{r}) > U(\mathbf{r})$, and therefore $\alpha > 0$.

Turning to the contribution $\boldsymbol{\rho}' = \pm a$, we assume simple cubic symmetry of the lattice and combine the contributions of the nearest neighbors at the distances $\pm a$ in x-, y-, and z-direction, respectively. In this way we find

$$\varepsilon_k = \varepsilon_O - \alpha - 2\beta\left(\cos k_x a + \cos k_y a + \cos k_z a\right) \tag{4.10}$$

With

$$-\beta = \int d\tau \, \varphi_o^*(\mathbf{r} - \mathbf{a})[U(\mathbf{r}) - U_a(\mathbf{r})]\varphi_O(\mathbf{r}) \tag{4.11}$$

The factor β contains the interaction between the nearest neighbors. In general, we have again $\beta > 0$.

Because of the interaction between each atom or molecule and its neighbors within the crystal lattice, the discrete energy levels split up and broaden into energy bands. From (4.10) we see, that the lowest energy value is $\varepsilon = \varepsilon_O - \alpha - 6\beta$. From the highest energy value $\varepsilon = \varepsilon_O - \alpha + 6\beta$ we see, that the width $\Delta\varepsilon$ of the energy band amounts to $\Delta\varepsilon = 12 \, \beta$. Since the overlap between $\varphi_o^*(\mathbf{r} - \mathbf{a})$ and $\varphi_O(\mathbf{r})$ increases with decreasing lattice constant a, then also the quantity β and, hence, the bandwidth $\Delta\varepsilon$ increases. In Fig. 4.3 we show this behavior schematically in the case of a one-dimensional chain.

So far we have considered only the interaction of an atom or molecule at a certain lattice site with its nearest neighbors. In order to achieve a higher accuracy

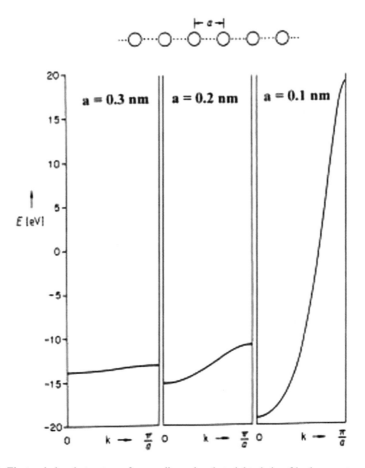

Fig. 4.3 Electronic band structure of a one-dimensional straight chain of hydrogen atoms with a distance between neighbors of 0.3, 0.2, and 0.1 nm, respectively. The electron energy ε is plotted as a function of the wave vector k. For a single isolated hydrogen atom the electron energy is −13.6 eV. The broadening of this energy level results in the energy bands. The energy width of these bands increases strongly with decreasing distance between neighboring atoms. (R. Hoffmann)

of the approximation, one can include also the interaction with the next-nearest neighbors (and perhaps beyond those).

At the end of our discussion of the bound-state approximation we note, that the interaction with the atoms or molecules at the neighboring lattice sites results in a broadening of the discrete energy values, associated with an isolated atom at a lattice site, into energy bands, the bandwidth of which increases with decreasing distance between the nearest neighbors.

4.2 Nearly-Free Electron Approximation (Rudolf Peierls)

In the other important limiting case, the approximation with free electrons, discussed for the first time by Rudolf E. Peierls also in Leipzig, nearly-free electrons are assumed, and the perturbation by the periodic potential of the crystal lattice is considered to be small. In this case the electrons can propagate freely through the crystal in the form of matter waves. However, this free propagation is interrupted if the matter waves undergo Bragg reflection at the crystal lattice.

If we completely ignore the potential energy $U(\mathbf{r})$ in the Schrödinger equation (4.1), we obtain as solution the wave function of free electrons:

$$\psi(\mathbf{r}) = e^{i\mathbf{k}\mathbf{r}}. \tag{4.12}$$

By inserting (4.12) into (4.1), in the case $U = 0$ one finds the electron energy

$$\varepsilon = \hbar^2 k^2 / 2m. \tag{4.13}$$

The wave function (4.12) represents a plane wave with the wave vector $\mathbf{k}$. However, a small periodic potential $U(\mathbf{r})$ in (4.1) causes a strong effect, if the wave vector $\mathbf{k}$ lies close to the boundary of a Brillouin zone. At the zone boundary, the wave vector $\mathbf{k}$ exactly satisfies the Bragg condition (2.10), such that the wave experiences Bragg reflection. As an example, schematically indicated in Fig. 4.4, we take the point $\mathbf{G_1}$ of the reciprocal lattice and the zone boundary at $\mathbf{G_1}/2$. Due to the Bragg reflection, the wave vector $\mathbf{k} = \mathbf{G_1}/2$ changes from $\mathbf{k}$ to $\mathbf{k}' = \mathbf{k} - \mathbf{G_1} = -\mathbf{G_1}/2$.

In the presence of the periodic potential $U(\mathbf{r})$, the solution of the Schrödinger equation can be written as superposition of plane waves:

$$\psi(\mathbf{r}) = \sum_{\mathbf{k}} c_{\mathbf{k}} e^{i\mathbf{k}\mathbf{r}}. \tag{4.14}$$

If we express the small periodic potential $U(\mathbf{r})$ of the crystal lattice in the form of (2.7), then in the case $\mathbf{k} = \mathbf{G_1}/2$ (after some calculations) one obtains the electron energy

Fig. 4.4 Bragg reflection from $\mathbf{k} = \mathbf{G_1}/2$ to $\mathbf{k}' = \mathbf{k} - \mathbf{G_1} = -\mathbf{G_1}/2$

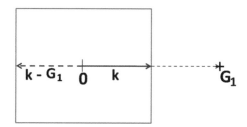

$$\varepsilon = \varepsilon_k \pm u_{\mathbf{G}_1}. \tag{4.15}$$

Between the energies $\varepsilon_k + u_{\mathbf{G}_1}$ and $\varepsilon_k - u_{\mathbf{G}_1}$ there appears a forbidden energy gap, in which the solution (4.14) of a propagating wave does not exist. The spectral energy curve $\varepsilon(\mathbf{k})$ always approaches the boundaries of the Brillouin zones with zero slope. The magnitude of the energy gap increases with increasing coefficient $u_{\mathbf{G}_1}$ in (2.7), i.e., with increasing potential energy $U(\mathbf{r})$ of the crystal lattice. In Fig. 4.5 we show the appearance of the forbidden energy gaps due to the Bragg reflection at the periodic potential of the crystal and the comparison with the case of the perfectly free electrons.

In 1931 Allan H. Wilson from England had joined Heisenberg's group in Leipzig, and finally it was he who provided the definite answer to the question, of whether a crystal is an electrical conductor, a semiconductor, or an insulator (Fig. 4.6). According to his proposal, which then turned out to be correct, the energy bands of the electrons in a crystal are responsible for the differences in the electrical conductivity. Here the decisive argument is based on the fact, that the preferential motion of the electrons along a certain direction (of an applied electric field) is only possible, if the electronic states of the energy spectrum relevant under the non-equilibrium exist and can be occupied. If a band is only partly filled with electrons we have metallic electrical conductivity. On the other hand, one obtains an electrical insulator, if all energy bands are completely filled with electrons, and if at the same time no empty band exists nearby along the energy axis. Any band which is completely filled cannot contribute to the electrical conductivity, since the distribution of the velocities of all electrons in the band cannot be changed. However, in order to conduct an electrical current, the velocities of the electrons

Fig. 4.5 Energy bands in crystals. In the crystal lattice the continuous energy spectrum of free electrons (**a**) is divided into different energy bands, which are separated from each other by forbidden energy gaps (**b**). E energy; k wave vector; a distance between neighbors in the crystal lattice

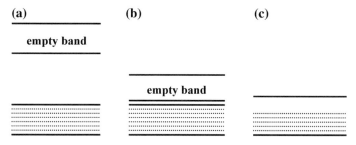

Fig. 4.6 Energy-band model of the electric conductivity of crystals. **a** A completely filled energy band having a large energy distance ("energy gap") to the next higher, but still completely empty energy band, yields an electric insulator. **b** A completely filled energy band with only a small energy distance to the next higher, but still nearly empty energy band, results in a semiconductor. **c** A well, but not yet completely filled energy band yields the electric conductance of a metal

must be redistributed in favor of the flow of current, which is impossible for a completely filled band, since unoccupied energy values are not available. There then remains as an interesting situation, the case where an empty band exists energetically close on top of a filled band, such that electrons can be transferred from the lower to the upper band by means of their thermal excitation energy. Hence, the energy gap between both bands must be sufficiently small. In this case we deal with a semiconductor, a subject we will discuss in Chap. 6. Typical values of the energy gap are about 0.1–1 eV in the case of semiconductors and about 10 eV in the case of insulators.

For the theoretical physicists in Leipzig it has taken only three years to solve the problem of the electrical conductivity of crystals in terms of the energy bands of the electrons.

Chapter 5
Metals Obey the Rules of Quantum Statistics

Abstract The classical Drude-Lorentz model could explain some aspects of the electronic properties of metals such as the Wiedemann-Franz law. However, quantum statistics and the Fermi distribution of the electron energies turned into the key for the electronic theory of metals, including the important concept of the Fermi surface in momentum space. Over a wide temperature range the electronic transport properties of metals are dominated by the interaction between electrons and phonons (Bloch-Grüneisen law). Thermoelectricity (Peltier and Seebeck effect) is discussed briefly.

5.1 Drude-Lorentz Model

Before the quantum mechanical foundations discussed in Chap. 4 were developed, there already existed classical models for describing the behavior of electrons in metals. Here the dominating model was due to Paul Drude and Hendrik Antoon Lorentz. The electrons in a metal were assumed to represent an ideal gas which can move freely within the crystal lattice. Further, only one kind of mobile carriers of negative electric charge was assumed to exist. In some way, the presence of the atoms of the crystal lattice was ignored. However, they should occasionally result in collisions with the mobile electrons. In this way one could arrive at a finite value of the "electron mean free path" and, hence, a finite electrical conductivity.

In an electric field $\mathbf{E}$ the electrons in a crystal experience the force

$$\mathbf{F} = \hbar \frac{d\mathbf{k}}{dt} = -e\,\mathbf{E}. \tag{5.1}$$

Here $\mathbf{k}$ denotes the wave vector and e the charge of the electrons. The quantity $\hbar\mathbf{k}$ is the mechanical momentum of the electrons. According to (5.1), after the time Δt the wave vector increases by the amount $\Delta\mathbf{k}$. In the absence of any scattering processes, the increment $\Delta\mathbf{k}$ will grow further and further. However, since electrons

© Springer Nature Switzerland AG 2019
R. P. Huebener, *Conductors, Semiconductors,*
Superconductors, Undergraduate Lecture Notes in Physics,
https://doi.org/10.1007/978-3-030-31420-0_5

always experience scattering processes (due to phonons or defects in the crystal lattice), Equation (5.1) is valid only during a limited time interval. This time interval is the average time τ between two scattering events, and one finds

$$\hbar \, \Delta \mathbf{k} = \mathbf{F} \, \tau = \mathrm{m} \, \Delta \mathbf{v}. \tag{5.2}$$

Here m denotes the mass and $\Delta \mathbf{v}$ the drift velocity of the electrons. Together with (5.1) one obtains for the electric current density $\mathbf{j}$:

$$\mathbf{j} = \mathrm{n}(-\mathrm{e})\Delta \mathbf{v} = \mathrm{ne}^2 \, \frac{\tau}{m} \, \mathbf{E}, \tag{5.3}$$

where n denotes the density of the electrons. With the electrical conductivity σ from the relation $\mathbf{j} = \sigma \, \mathbf{E}$ one finds:

$$\sigma = \mathrm{n} \, \mathrm{e}^2 \tau / \mathrm{m}. \tag{5.4}$$

As the result we obtain Ohm's law, based on the average "relaxation time" τ, assumed to be independent of the electric field $\mathbf{E}$.

Since electrons also carry heat energy in addition to their electric charge, they also contribute to the heat conductivity of metals. This contribution of the electrons represents a second important mechanism of heat conduction, which must be taken into account in metals, in addition to the heat transport by the phonons, which we have discussed in Chap. 3. Often the contributions from both mechanisms are of similar magnitude. Since the transport of heat energy and of electric charge is due to the same electrons, one expects that the electronic part of the heat conductivity and the electrical conductivity will be proportional to each other, in agreement with experiment. This can be demonstrated using the following argument. In a temperature gradient dT/dx an energy current of density w flows from the hot to the cold side, given by

$$\mathrm{w} = (1/2)\mathrm{n} \, \mathrm{v} \, \{\varepsilon(\mathrm{T}[\mathrm{x}-\mathrm{v}\tau]) - \varepsilon(\mathrm{T}[\mathrm{x}+\mathrm{v}\tau])\} \tag{5.5a}$$

$$\mathrm{w} = \mathrm{n} \, \mathrm{v}^2 \tau \frac{d\varepsilon}{dT} \left(-\frac{dT}{dx} \right) \tag{5.5b}$$

Here, ε denotes the particle energy. In Fig. (5.1) we illustrate the argument leading to (5.5a, 5.5b).

The mean free path $\ell = \mathrm{v} \, \tau$ of the electrons represents the relevant length scale, at which the average motion of the electrons is affected along the temperature gradient. Based on the average energy $\frac{1}{2} \, k_B T$ per degree of freedom of a free particle, in the case of three degrees of freedom the energy ε is given by

$$\varepsilon = \frac{1}{2} \mathrm{m} \, \mathrm{v}^2 = \frac{3}{2} k_B T \tag{5.6}$$

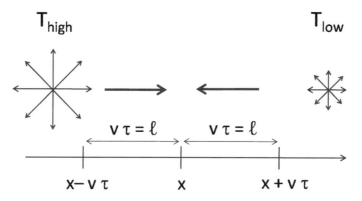

Fig. 5.1 Illustration regarding the heat current density of (5.5a, 5.5b). The arrows near T_{high} and T_{low} indicate the average electron energy

and using

$$\frac{d\varepsilon}{dT} = C_V = \frac{3}{2}k_B \qquad (5.7)$$

we obtain the energy current density

$$w = -n\left(\frac{3}{2}k_B\right)^2 \frac{\tau T}{m}\frac{dT}{dx} = -\kappa_e \frac{dT}{dx}. \qquad (5.8)$$

In (5.8) we have introduced the heat conductivity κ_e of the electrons. Together with (5.4) we find

$$L = \frac{\kappa_e}{\sigma T} = \frac{9}{4}(k_B/e)^2. \qquad (5.9)$$

The ratio L is referred to as the Lorenz number, named after Ludwig Lorenz. (A more accurate averaging yields the prefactor $\pi^2/3$ instead of 9/4). The result in (5.9) is the famous Wiedemann-Franz law. Often this law is very useful for estimating the heat conductivity of the electrons in a metal, if the electrical conductivity, which can be measured relatively easily, is known.

The explanation of the experimentally observed Wiedemann-Franz law was one of the successes of the Drude-Lorentz model. However, the model failed to predict the electrical and the thermal conductivity of the electrons separately, i.e., not just the ratio between both conductivities. But further and much more fundamental difficulties appeared with respect to the specific heat of the electrons. Their contribution to the specific heat was found to be much smaller than expected from classical concepts. Again, the solution of this problem was provided by the new quantum mechanics and in particular by the application of the "Pauli principle".

5.2   Quantum Statistics, Fermi Distribution

Again, it is the exact identity of the electrons as elementary particles, which makes them indistinguishable and which requires a new form of quantum statistics in order to obtain the probability distribution of the states. We have noted this already for the quantized lattice vibrations in terms of bosons. The quantum statistics deviate strongly from the classical (Maxwell-Boltzmann-) statistics, as we have discussed in Chap. 3 in the case of the Bose-Einstein distribution (3.2) of particles having vanishing or integer angular momentum (phonons and photons). However, now we deal with the electrons as elementary particles, having a half-integer angular momentum. In 1925 Wolfgang Pauli had formulated his famous exclusion principle, which states that each quantum mechanical state of a system can at most be occupied by a single electron. Here the important point is that an electron carries a half-integer angular momentum. In this way Pauli was able to explain the closure of the electron shells of the atoms. In 1926 Enrico Fermi from Italy and Paul Adrien Maurice Dirac showed independently from each other, that the application of the Pauli principle also leads to a new form of quantum statistics which today are referred to as Fermi-Dirac statistics. In general, Fermi-Dirac statistics are valid for elementary particles with a half-integer angular momentum, as is the case for the electrons. Such particles are referred to as fermions. Their angular momentum is also called spin. Because of the quantization of the direction of the angular momentum, the half-integer spin of the electrons can be oriented along only two possible directions. According to the Pauli principle, each state can be occupied for each of the two spin directions at most only by a single electron. Therefore, the many electrons in a metal must distribute themselves over many states with different energies within an energy band. In this way the electrons in an energy band must occupy sequentially the different "seats" with increasing energy. The last electron then must take the highest energy level. This highest energy level of the occupied states is referred to as the Fermi energy and the corresponding energy distribution of the electrons as the Fermi distribution (or Fermi-Dirac distribution). In the following we denote the Fermi energy by ε_F.

The Fermi-Dirac distribution function $f(\varepsilon)$ is given by

$$f(\varepsilon) = \frac{1}{e^{(\varepsilon - \varepsilon_F)/k_B T} + 1}. \tag{5.10}$$

ε is the particle energy and ε_F the Fermi energy (or the chemical potential). The function $f(\varepsilon)$ is shown in Fig. 5.2.

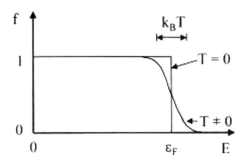

Fig. 5.2 Fermi distribution function: Because of the Pauli exclusion principle each quantum mechanical state in a crystal can be occupied at most by only one electron. At the temperature of zero Kelvin, for all energies up to the Fermi energy ε_F the Fermi distribution function has the value of one, and at ε_F it drops abruptly from one to zero. At a finite temperature T, the drop of the Fermi distribution function from one to zero is smeared out along the energy axis, and it occurs at the Fermi energy within an energy width of about $k_B T$

Mathematically the Fermi distribution $f(\varepsilon)$ is a simple function of the electron energy. Between zero energy and the Fermi energy this function has a value of one, since in this energy interval the states can be occupied only by a single electron. At the Fermi energy the function abruptly drops from one to zero, having approximately the shape of a rectangle. However, this rectangular shape is exactly valid only at zero temperature. At a temperature T, different from zero, the drop of the Fermi function from one to zero is smeared out along the energy axis and happens within an energy interval of about $k_B T$ at the Fermi energy (Fig. 5.2). This energy width $k_B T$ has already appeared in our discussion of the thermal energy associated with the individual degrees of freedom of the normal modes of the crystal lattice.

Now we return again to the specific heat of the electrons. Since the electrons in a metal must obey a highly restrictive prescription in the form of the Fermi distribution, nearly all electrons in the relevant energy band are energetically fixed and cannot change their energetic state. If we arbitrarily pick out an electron, we see that all energetically neighboring states are already occupied. Hence, this electron can reach an unoccupied state only by means of a very large jump in energy, effecting a physical change in this way. However, in general such a large energy jump is impossible. Only for the few electrons near the Fermi energy is this possible, since they are energetically sufficiently close to states which are still unoccupied and which they can reach by thermal excitation. The fraction of the electrons in this exceptional energy interval is approximately given by $k_B T/\varepsilon_F$. It is also exactly this fraction, which contributes to the specific heat of the electrons. We see that the specific heat in a metal must be reduced by this factor $k_B T/\varepsilon_F$ compared with the value expected from the classical theory and that, furthermore, it becomes proportional to the absolute temperature. Both results agree well with experiment. In this way it was quantum statistics which again removed the difficulties of the classical theories with the specific heat, in this case for the electrons. Using the relevant numbers for the monovalent metals, one finds for room temperature approximately the value $k_B T/\varepsilon_F = 0.01$. Compared with the Fermi energy, the transition from the occupied to the unoccupied states happens within a relatively narrow energy interval. Hence, the rectangular shape of the Fermi function mentioned above is still reasonably well preserved.

Exactly the same argument as we have used above for the specific heat of the electrons also applies to the paramagnetism of the electrons in metals. This was pointed out for the first time by Wolfgang Pauli, who explained in this way the relatively small value of the "paramagnetic susceptibility" in metals and its independence of the temperature. We will return to this subject in Chap. 10.

5.3   Fermi Surface

At this stage we recall that the states of the electrons in a crystal represent planar waves, as formulated for the first time by Felix Bloch and Rudolf E. Peierls in their quantum mechanical theory. The planar waves are characterized by their wavelength and their propagation direction. Both qualities are combined in the "wave vector" **k**. Its direction indicates the propagation direction of the wave. Its absolute value k, referred to as the wave number, is equal to the inverse of the wavelength λ except for the factor 2π: $k = 2\pi/\lambda$. The state of the electrons in the crystal is not unequivocally identified by the energy alone. For the same electron energy the wave vectors of the matter waves can still point in all different directions in the crystal, in this way defining different states in view of the Pauli principle. Hence, the Fermi distribution of the electrons must apply to all directions of the wave vectors separately. With increasing energy of the electrons, the magnitude of their wave vectors also increases. Hence, the Fermi energy as the maximum energy of the occupied states also corresponds to a maximum value of the wave vector. This maximum value is referred to as the Fermi wave vector $\mathbf{k_F}$, and all states up to $\mathbf{k_F}$ are occupied by electrons. All states above $\mathbf{k_F}$ remain unoccupied. From this discussion we see that our treatment of the quantum mechanical electron states in a crystal must be extended to a three-dimensional space of the wave vectors, the so-called **k**-space. Since the wave vector is proportional to the particle momentum, this space is also referred to as the momentum space. In this momentum space the Fermi wave vectors $\mathbf{k_F}$ with their magnitudes and their directions represent a surface, the "Fermi surface", for the particular material. The Fermi surface is one of the most important concepts for the discussion of the electronic crystal properties (Fig. 5.3). The Fermi energy ε_F and the Fermi wave vector $\mathbf{k_F}$ are connected by the relation

$$\varepsilon_F = \hbar^2 k_F^2 / 2\,m \tag{5.11}$$

This relation is shown in Fig. 5.4.

In the simplest case, if the influence of the crystal lattice on the de Broglie waves associated with the electrons is negligible, the Fermi surface has a spherical shape. We deal with this case to a good approximation in the monovalent metals, as for

Fig. 5.3 Fermi surface of copper. – The Fermi surface exists in momentum space, and it indicates up to which value all wave vectors are used up in order to accommodate the available electrons. In the simplest case the Fermi surface is a sphere, the radius of which is given by the Fermi wave vector k_F. The picture shows the Fermi surface of copper for which, historically, deviations from the spherical shape were detected for the first time. The short "necks" visible in eight different directions are a characteristic feature of the Fermi surface of copper

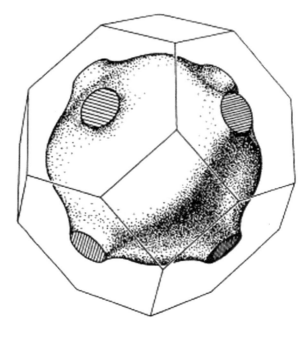

Fig. 5.4 The energy spectrum $\varepsilon(\mathbf{k})$ of the electrons is occupied up to the Fermi energy ε_F and up to the Fermi wave vector k_F. (It is assumed that $k_B T \ll \varepsilon_F$)

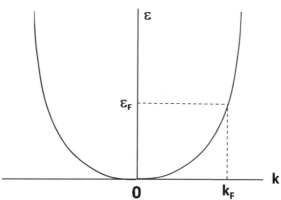

example in the alkali metals. As we have discussed already in Chap. 4, the perturbation arising from the crystal lattice becomes stronger as the electron energy approaches the values which satisfy the condition of Bragg reflection (at the boundary of a Brillouin zone). If the Fermi energy gets close to these values, the Fermi surface deviates appreciably from the spherical shape and displays a characteristic anisotropy in momentum space. We encounter such a case in particular in the multivalent metals, as for example in aluminum.

A distinct anisotropy of the Fermi surface was experimentally observed for the first time by the Englishman Alfred Brian Pippard in measurements of the microwave surface resistance of copper (Fig. 5.3). During the Second World War,

Pippard had participated in England in the development of radar technology, which turned out to become the key defense determining the outcome of the Battle of Britain. (It is also said, that radar technology had won the Second World War). For the young Pippard his experience with the new microwave technique had been the reason for taking up a thesis subject dealing with microwaves. Pippard performed his crucial measurements during the academic year 1955/1956 as a guest scientist at the Institute for the Study of Metals in Chicago, since in this Institute single crystals of copper with highly polished surfaces could be prepared better than anywhere else. This first experimental observation of the anisotropy of the Fermi surface in a metal by Pippard immediately triggered considerable research activities on the subject of the Fermi surface in many laboratories, which subsequently lasted for many years. Within this context experiments in high magnetic fields and, in particular, the "de Haas–van Alphen effect" quickly gained much importance. This will be discussed in Chap. 7.

An extremely fruitful comment starting the application of the geometric concept of the Fermi surface in momentum space originated from the Norwegian Lars Onsager, during a visit to Cambridge, England in the early 1950s. The pioneering research in the field of Fermi surfaces in metals by David Shoenberg and his coworkers in the Royal Society Mond Laboratory greatly benefitted from Onsager's remark. Only this concept gradually made it possible to interpret correctly the many experimental data for metals. As a main result, it had become clear that many properties of metals are determined only by a small fraction of the electrons residing in close proximity to the Fermi surface. Here the key role of two-dimensional interfaces, this time in momentum space, is impressively demonstrated again, in some way similar to the appearance of all life on the surface of the earth in biology.

5.4 Bloch-Grüneisen Law

In his dissertation Bloch had also developed a theory of the electrical resistance of metals. As the key point he treated the collision processes of the electrons with the vibrational quanta of the crystal lattice. Here he took into account that, during such a collision process, the electrons can exchange energy with the crystal lattice in the form of individual phonons. As a final result, Bloch found the famous Bloch-Grüneisen law for the temperature dependence of the electrical resistance in metals. In this law the "Debye temperature" θ plays a role. At the temperature θ the thermal energy $k_B\theta$ is equal to the Debye energy $\hbar\omega_D$, which we have discussed before as the maximum value in the energy spectrum of the phonons: $k_B\theta = \hbar\omega_D$. At temperatures above the Debye temperature θ ($T \gg \theta$) all phonon states up to the Debye frequency ω_D are occupied, and the number of phonons per state is proportional to T (as we see from the distribution (3.2)). Therefore, in this temperature range we expect for the electrical resistivity ρ that $\rho \sim T$. On the other hand, at temperatures much smaller than the Debye temperature θ ($T \ll \theta$) the number of occupied phonon states (which contribute to the scattering rate $\tau^{-1)}$

Fig. 5.5 Scattering by the
angle δ reduces the drift
velocity v_D in forward
direction to $v_D \cos \delta$

forward direction

increases proportional to T^3, as we have seen in connection with (3.4). An additional factor proportional to T^2 results from the temperature dependent magnitude of the scattering angle δ.

This is illustrated in Fig. 5.5. Denoting the drift velocity of the electrons by v_D, after scattering by the angle δ, the drift velocity in forward direction is $v_D - \Delta v_D = v_D \cos \delta$. The lost relative drift velocity in forward direction is $\Delta v_D / v_D = 1 - \cos \delta \approx \delta^2$, assuming the scattering angle being small. In the case $T \ll \theta$ the scattering angle δ shows the proportionality $\delta \sim$ phonon wave vector $K \sim T$. The loss of drift velocity of the electrons along the preferential direction increases proportional to δ^2 and, hence, proportional to T^2. Therefore, in this temperature range we finally obtain $\rho \sim T^5$. This overall behavior, $\rho \sim T^5$ at $T \ll \theta$ and $\rho \sim T$ at $T \gg \theta$, represents the famous Bloch-Grüneisen law of the electric resistance. This law has been well confirmed experimentally (Fig. 5.6).

The Bloch-Grüneisen law for the temperature dependence of the electrical resistance only takes into account the collisions between the electrons and the quantized vibrations of the crystal lattice. The variation in the resistance with

Fig. 5.6 Bloch-Grüneisen
law for the temperature
dependence of the electrical
resistance of different metals,
the Debye temperature (here
denoted by T_D) of which is
indicated in Kelvin. The
temperature is given in units
of the Debye temperature T_D
and the electric resistance in
units of its value $R(T_D)$ at the
Debye temperature. For the
different metals one obtains a
universal curve.
(W. Meissner)

temperature expressed by this law is mainly due to the decreasing number of phonons with decreasing temperature. However, in the crystal the electrons also experience other collision processes limiting the electrical conductivity. In this context structural lattice defects or chemical impurities, perturbing the perfectly regular periodic lattice structure of the crystal, play a major role. Alloys also represent an important example, where collision processes other than those with phonons are important. These defects in the crystal lattice likewise contribute to the electrical resistance. In general, the scattering rates τ_i^{-1} of different mechanisms, for example, the collisions of the electrons with the phonons and with the defects in the crystal lattice, can simply be added, and we obtain for the total scattering rate τ^{-1}:

$$\frac{1}{\tau} = \sum_i \frac{1}{\tau_i} \tag{5.12}$$

This quality of the additivity of the different mechanisms contributing to the resistance is referred to as Mathiessen's law. Since the contribution of the phonons strongly decreases with decreasing temperature, at sufficiently low temperatures only the contribution of the lattice defects, the "residual resistance", remains. This contribution is nearly independent of temperature. The magnitude of this residual resistance provides an easily accessible indication of the purity level of the metal. In highly pure metals this residual resistance is a few hundred up to a few thousand times smaller than the resistance at room temperature, the latter being dominated by the contribution of the phonons.

5.5 Thermoelectricity

In our discussion of the electrical conductivity and of the thermal conductivity of metals we had to deal only with a single external influence acting on the metal. In the former case we are concerned with a gradient in the electrical potential due to an electric field and in the latter case with a gradient in the temperature. However, it is also possible that both external influences act simultaneously. In this case we deal with the thermoelectric phenomena, which we will discuss next.

We start with the Peltier effect. It results from the fact that an electric current always transports the heat energy of the moving charge carriers along with their electric charge. If two electrical conductors from different materials are connected in series, the heat current can pile up at the location of the joint, if the heat current carried by the same electric current is different in the two materials. Depending upon the direction of the current, a heating effect or a cooling effect appears at the joint (Fig. 5.7a). This effect is named after the Frenchman Jean Charles Athanase Peltier, who discovered it in 1834. The Peltier coefficient Π is defined as follows:

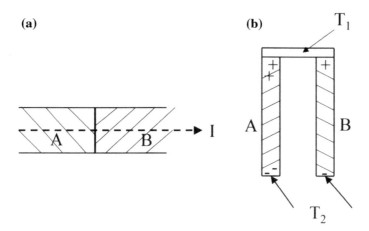

Fig. 5.7 **a** Peltier effect: If the electric current I passes through the contact zone between two different metals A and B, having a different value of the heat current carried by the same electric current, heating or cooling of the contact zone results, depending upon the current direction. **b** Seebeck effect: In the temperature gradient between the higher temperature T_1 (upper side) and the lower temperature T_2 (lower side) in an electrical conductor the mobile electrons are transported from the hot to the cold end of the conductor by means of thermal diffusion. Therefore, at both ends of the conductor, electric charges of opposite sign, respectively, accumulate. The direction of this thermal diffusion process is determined by the details of the Fermi surface and of the collision processes of the electrons. The figure shows two different metals A and B, which are soldered together at the end with the higher temperature. The difference in the thermal diffusion between the two metals results in an electric voltage, the "thermoelectric voltage", between the lower ends of the two metals

$$\Pi \equiv \frac{\text{heat current density } w_x}{\text{electric current density } j_x} \tag{5.13}$$

At the location of the joint between conductors A and B, the delivered or absorbed Peltier heat per unit time and cross-sectional area amounts to $(\Pi_A - \Pi_B)j_x$. This leads to heating or cooling at the joint, depending upon the current direction. However, in metals, the Peltier effect is relatively small and, hence, it is not interesting for applications in cryogenics. Again, the small value of the Peltier effect in metals results from the severe restriction imposed by quantum statistics, which allows only the small fraction $k_B T/\varepsilon_F$ of all electrons to participate in the transport phenomena, in close similarity to what we have seen before, for the specific heat of the electrons.

Next we turn to the Seebeck effect as the second thermoelectric phenomenon (Fig. 5.7b). It was observed for the first time by the German Thomas Johann Seebeck in 1821. Seebeck was born in Reval (today Tallinn). After he had studied medicine in Berlin and Göttingen, he practiced as a medical doctor in Göttingen. However, subsequently he turned to physical research and worked as a private scholar in Jena, Bayreuth, and Nuremberg. The Seebeck effect represents a special case of the "thermal diffusion" of particles in a temperature gradient dT/dx. We are well familiar with this phenomenon: the deposition of the relatively heavy dust particles from the air on the bright wall paper of a cold wall immediately behind a heating pipeline is caused by thermal diffusion. This same phenomenon is the underlying principle of the Clusius separation column utilized for isotope separation. The German physico-chemist Klaus Clusius invented the separation column process in 1938. Subsequently this process has played a highly prominent role in the American Manhattan Project during the Second World War for the enrichment of the Uranium isotope ^{235}U. At the time in Oak Ridge in the Federal State of Tennessee, a huge plant had been constructed consisting of 2142 separation columns, each of 16 m in height. Along the axis of each tube a thin nickel tube was

Fig. 5.8 Thermal diffusion: A small part of the thermal-diffusion plant constructed during the Manhattan Project in Oak Ridge in the American Federal State of Tennessee. We can see the separation columns for the enrichment of the uranium isotope ^{235}U based on the principle invented by Klaus Clusius. (*Photo* AEC photo of Ed Westcott)

placed, which was heated and surrounded by a larger copper tube. The Uranium was fed into the plant in the form of gaseous UF_6 (Fig. 5.8).

Thermal diffusion is caused by the thermal force - S_{tr} dT/dx acting on the particles in a temperature gradient dT/dx. The quantity S_{tr} is the "transport entropy" per particle. In the case of the Seebeck effect, thermal diffusion leads to the accumulation of electric charges of opposite sign at the two ends of the conductor, respectively. As a result, an electric field E_x is generated. Under equilibrium, the thermal force is compensated by the electrostatic force - e E_x (we use - e for the charge), and we obtain the following equation for the forces:

$$-S_{tr} \frac{dT}{dx} = -eE_x = e\frac{dU}{dx}. \tag{5.14}$$

Here U denotes the electric potential. From (5.14) one obtains the Seebeck coefficient (thermopower) S:

$$S \equiv \frac{\Delta U}{\Delta T} = -\frac{S_{tr}}{e}. \tag{5.15}$$

The thermoelectric voltage is always measured between two materials relative to each other (Fig. 5.7b). In the case of two electric conductors A and B, which are soldered together at one end, the temperature difference ΔT between the two ends yields the electric potential difference $\Delta U = (S_A - S_B) \Delta T$. The thermoelectric voltage ΔU is proportional to the temperature difference ΔT between both ends of the electric conductors. Therefore, it is well suited for measurements of the temperature, if the temperature of one end is exactly known. In the form of the "thermoelements" consisting of electrically conducting wires of two different metals or alloys, the Seebeck effect is often used for thermometry.

From transport theory one can derive mathematical formulas for the thermoelectric coefficients, which incorporate details of the scattering processes of the electrons and of the geometry of the Fermi surface. Here we do not want to pursue this any further. However, one should point out, that in metals the reduction factor $k_B T/\varepsilon_F$ applies again because of the Fermi distribution, similar to the case of the specific heat of the electrons.

At the end of this section on thermoelectricity we briefly address a subject referred to as "phonon drag". We explain this phenomenon in the case of the Seebeck effect. In a temperature gradient, a phonon current transporting heat is generated, in addition to the thermal diffusion of the electrons, as we have discussed above in the context of the heat conductivity of metals. Due to the electron-phonon interaction, this phonon current drags the electrons along, thereby causing the "phonon-drag component" of the Seebeck coefficient. The other component, remaining if this phonon current would be absent, is referred to as the "electron diffusion component". The phonon-drag component of the Seebeck coefficient varies with temperature similarly as the heat conductivity of the crystal lattice shown in Fig. 3.6. With increasing temperature it passes through a distinct maximum, and it vanishes generally above room temperature. (In the case of the electrical conductivity, the phonon-drag effect is negligible, since the electron-phonon interaction enters twice: first, the electrical current sets up a phonon current, which then acts back on the electrons. This represents an effect of second order).

The Peltier coefficient Π and the Seebeck coefficient S of a material are connected with each other through the Thomson relation

$$\Pi = TS \qquad\qquad (5.16)$$

This relation between both coefficients represents a prominent example of the famous reciprocity scheme of Lars Onsager for transport coefficients.

The rapid development of the theory of metals, which we have briefly summarized above, is also clearly apparent from the much more detailed treatments published soon after the foundations of the new quantum mechanics were established. Here we mention in particular the book "The Theory of the Properties of Metals and Alloys" by Nevill Francis Mott and H. Jones, as well as the book "The Theory of Metals" by Allan H. Wilson, both books dating from the year 1936. An impressive milestone is a large review article "Electron Theory of Metals" (German title: "Elektronentheorie der Metalle") by Arnold Sommerfeld and Hans Bethe for the German Handbook of Physics from 1933. By far the largest part of this review article had been written by Bethe at the young age of only 27 years. Even today this book-sized article continues to be relevant and highly useful. Already in 1931 the French Léon Brillouin had summarized the status of the field in his book "Quantum Statistics and its Application to the Electron Theory of Metals" (German title: "Die Quantenstatistik und ihre Anwendung auf die Elektronentheorie der Metalle").

The severe restriction imposed upon the electrons as fermions by quantum statistics is the main idea in this chapter about the properties of metals. However, it can also happen that two electrons combine forming a pair, but no longer having the half-integer angular momentum of a fermion as a pair. Instead, for example, the pair may have zero total angular momentum. A total spin with the value of zero results, if the individual spins of both partners are oriented in opposite directions to each other. In this case the Pauli Principle is no longer valid, and many electron pairs can occupy the same quantum mechanical state. In Chap. 8 we will discuss exactly how this happens in the phenomenon of superconductivity.

Chapter 6
Less Can Be More: Semiconductors

Abstract If the electronic band structure allows the thermal excitation of electrons from the top of a completely filled valence band to the bottom of an empty conduction band, we deal with a semiconductor, such as germanium or silicon, or the compound semiconductors consisting of elements of the third and the fifth column of the Periodic Table (or of the second and the sixth column). The carefully controlled doping with donors and acceptors yields materials, which are extremely useful for many electronic applications. The metal-semiconductor contact and the rectification effect of the p-n junction are discussed. After its invention in 1947, the transistor underwent many evolutionary stages during its miniaturization. Our discussion includes photovoltaics, the light-emitting diode (LED), and the semiconductor laser.

Michael Faraday from England had already found in 1833, that the electrical resistance of silver-sulfide (Ag_2S) decreases with increasing temperature, whereas metals display the opposite temperature dependence. He observed a similar temperature dependence as in silver-sulfide also in a number of other substances, for which the electrical conductivity was always much smaller than for the typical metals. About 40 years later the German Ferdinand Braun was interested in the electrical conductance of galena (lead-sulfide) crystals and of other metal sulfides. He discovered that the electrical resistance in these materials depends even on the current direction. Such an effect had never been observed in metals. This effect was particularly clear if the electric current was injected into or extracted out of the substance on one side using a metallic needle. Braun had discovered the rectifying effect of a contact. Many years later this arrangement played a famous role for some time as a detector for radio waves. With his studies Braun widely opened the door for the subsequent investigation and utilization of a new class of electrical conductors: the "semiconductors". However, his most important studies, which he had carried out since 1895 as Professor of Physics at the University of Strassburg and for which he received the Nobel Prize for Physics in 1909 together with Guglielmo Marconi, were concerned with something else. In Strassburg, Braun developed the cathode ray tube, which became famous later on as Braun's tube. Among other things, it allows one to record high-frequency alternating currents with high time

© Springer Nature Switzerland AG 2019
R. P. Huebener, *Conductors, Semiconductors,*
Superconductors, Undergraduate Lecture Notes in Physics,
https://doi.org/10.1007/978-3-030-31420-0_6

resolution, and it is the central component of almost all television receivers until recently. (Presently, these apparatuses are replaced by flat screens.) Hence, it is not at all surprising, that under the direction of Ferdinand Braun the first lectures worldwide on high-frequency physics were given at the University of Strassburg in 1899.

In Chap. 4 we noted that the electrical conductance behavior of crystals is determined by the allowed energy bands and by the energy gaps between the bands of the energy spectrum of the electrons (Fig. 4.6). A conduction band only partly filled with electrons is the cause of the high electrical conductivity of metals, whereas a completely filled band does not contribute to the electrical conductivity. However, if an empty band is energetically located closely above a filled band, there exists an interesting new possibility. It is exactly this case which we have in semiconductors, and which already confronted Michael Faraday and Ferdinand Braun. If the energetic distance between the, at first empty, conduction band and the completely filled "valence band" located energetically underneath is sufficiently small, the electrons can jump across the small energy gap between both bands because of their thermal energy k_BT. In this way the conduction band can be populated with a relatively small number of electrons. These electrons in the conduction band then cause the electrical conductivity of the semiconductor. The number of electrons which can perform the energy jump from the valence band into the conduction band increases strongly with increasing temperature. Hence, the electrical conductivity of semiconductors also increases strongly with increasing temperature. Here semiconductors show exactly the opposite behavior to metals. As we have discussed in the Chap. 5, the electrical resistance of metals grows with increasing temperature, whereas the electrical conductance, as the inverse of the resistance, decreases. The electron concentration populating the conduction band in semiconductors because of the supply of the thermal energy k_BT is smaller by many orders of magnitude compared with a typical metal. Therefore, the electrical conductivity in semiconductors is also much smaller than in metals. In Fig. 6.1 we show the position of the valence band and of the conduction band along the vertical energy axis in the case of an intrinsic (undoped) semiconductor.

Fig. 6.1 Position of the valence band and of the conduction band along the energy axis in the case of an intrinsic (undoped) semiconductor

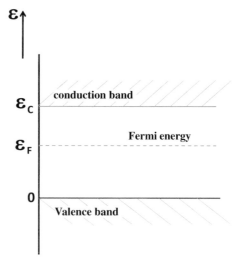

6.1   Intrinsic Semiconductors

The thermal excitation of electrons out of the valence band into the conduction band not only populates the conduction band from the bottom upwards with electrons, but in addition, because of the removal of electrons, the valence band becomes depopulated from the top downwards. Since electrons are now missing from near the upper edge of the valence band, one speaks about holes, (which are sometimes also referred to as defect electrons). Because of the existence of these holes, the electrons near the upper edge of the valence band can also participate in the electrical conduction mechanism, since the unoccupied states in this energy regime allow a change in the velocity distribution of the electrons required for the current flow. Now it is much more appropriate to describe the motion of the charge carriers near the upper edge of the valence band in terms of the dynamics of holes. A hole in the energy band of the negatively charged electrons corresponds exactly to a particle with the opposite, i.e., positive charge. The motion of a negatively charged electron, say, from left to right is completely equivalent to the motion of a positively charged hole from right to left. The useful and profound idea of the hole has been proposed for the first time for the physics of crystals by Werner Heisenberg. In a paper from 1931 dealing with the Pauli exclusion principle, Heisenberg first discusses the use of the wave equation of the holes in the context of the closed electron shells of an atom:

> If N denotes the number of electrons within the closed shell, …. it is shown that a Schrödinger equation for n electrons can be replaced also by a highly similar, equivalent Schrödinger equation for N − n holes (here in the English translation).

Heisenberg then discusses the use of the wave equation for the holes to explain the "anomalous Hall effect" in crystals. We will come back to the Hall effect in Chap. 7. From the Hall effect one can determine the sign of the moving charge carriers transporting the electric current. Many times the Hall effect had indicated a positive sign for the moving charge carriers, although it is the negatively charged electrons which make up the electric current flow in a conductor. Therefore, this lead to the notion of the anomalous Hall effect. In 1929, in Leipzig, Rudolf E. Peierls had proposed for the first time the correct interpretation of the anomalous Hall effect in terms of the appearance of holes in the occupation of the bands in the energy spectrum of the electrons.

In the case of the electrons, which are thermally excited from the valence band into the conduction band, we have $\varepsilon - \varepsilon_F \gg k_B T$, and the Fermi distribution (5.10) can be replaced by the Boltzmann distribution

$$f(\varepsilon) = e^{-(\varepsilon-\varepsilon_F)/k_B T} \qquad (6.1)$$

Using the energy scale shown in Fig. 6.1, one obtains for the electron concentration in the conduction band per volume:

$$n = \int_{\varepsilon_C}^{\infty} D_e(\varepsilon) f(\varepsilon) d\varepsilon. \tag{6.2}$$

Here

$$D_e(\varepsilon) = \frac{\sqrt{2}}{\pi^2} \left(\frac{m_e}{\hbar^2}\right)^{3/2} (\varepsilon - \varepsilon_C)^{1/2} \tag{6.3}$$

is the density of states of the electrons per volume; m_e is the electron mass. With (6.3) one obtains from (6.2):

$$n = \frac{\sqrt{2}}{\pi^2} \left(\frac{m_e}{\hbar^2}\right)^{3/2} e^{\varepsilon_F/k_B T} \int_{\varepsilon_C}^{\infty} (\varepsilon - \varepsilon_C)^{1/2} e^{-\varepsilon/k_B T} d\varepsilon \tag{6.4}$$

$$= 2(2\pi m_e k_B T/h^2)^{3/2} e^{-(\varepsilon_C - \varepsilon_F)/k_B T}. \tag{6.5}$$

At this point, we want to indicate the derivation of the density of states $D_e(\varepsilon)$ (6.3). We start with the number w(k) of the vibrations of the de-Broglie-waves of the electrons per k-interval (density of states) in **k**-space and, in this context, use periodic boundary conditions of the crystal (k = wave number of the electrons). We consider a crystal of length L along one coordinate axis. The periodic boundary condition requires, that the wave e^{ikx} has the same value at x = 0 and x = L, i.e., $e^{ikL} = 1$ or kL = n 2π, where n is an integer. The distance Δk between two subsequent k-values amounts to Δk = 2π/L. In the one-dimensional case we obtain $w_1(k) = 1/\Delta k = L/2\pi$. In the case of the three-dimensional **k**-space we have $w_3(\mathbf{k}) = (L/2\pi)^3$. One finds the density of states *per energy interval* (by considering a spherical shell of infinitesimal thickness ∂**k** in **k**-space)

$$D(\varepsilon) = w_3(\mathbf{k}) \frac{\partial(\mathbf{k} - \text{space} - \text{volume})}{\partial \varepsilon} = w_3(\mathbf{k}) \, 4\pi k^2 \frac{\partial k}{\partial \varepsilon}. \tag{6.6}$$

With the electron energy ε from (4.13), after a few steps one finally obtains

$$D(\varepsilon) = \frac{\sqrt{2}}{\pi^2} \left(\frac{m_e}{\hbar^2}\right)^{3/2} V \varepsilon^{1/2}. \tag{6.7}$$

Comparing (6.7) with expression (6.3), we note, that the latter is given per crystal volume, and that the conduction band is located above the energy ε_C.

The electrons transferred into the conduction band are now missing in the valence band, where they represent holes with the distribution function $f_h(\varepsilon) = 1 - f(\varepsilon)$. Analog to (6.5), the hole concentration per volume in the valence band is:

$$p = 2(2\pi \, m_h \, k_B \, T/h^2)^{3/2} e^{-\varepsilon_F/k_B T}. \tag{6.8}$$

Here, m_h is the mass of the holes. The product $n \, p$ is independent of the Fermi energy ε_F:

$$n \, p = 4(2\pi \, k_B T/h^2)^3 (m_e \, m_h)^{3/2} e^{-\varepsilon_C/k_B T}. \tag{6.9}$$

So far, we have considered only "intrinsic semiconductors". Therefore, we have:

$$n = p = 2(2\pi \, k_B T/h^2)^{3/2} (m_e \, m_h)^{3/4} e^{-\varepsilon_C/2 k_B T}. \tag{6.10}$$

From (6.5) and (6.8) one finds

$$m_e^{3/2} e^{-(\varepsilon_C - \varepsilon_F)/k_B T} = m_h^{3/2} e^{-\varepsilon_F/k_B T} \tag{6.11}$$

and from (6.11)

$$\varepsilon_F = \frac{\varepsilon_C}{2} + \frac{3}{4} k_B T \log(m_h/m_e). \tag{6.12}$$

In the case $m_h = m_e$ we find $\varepsilon_F = \varepsilon_C/2$, and the Fermi energy is located exactly in the middle of the energy gap.

The population of the conduction band with electrons by thermal activation and the simultaneous generation of holes near the upper edge of the valence band is a characteristic property of intrinsic semiconductors. Since the 1930s the study of these materials has grown steadily, a strong driving force being the possible technical applications. Initially the interest concentrated on copper-oxide and selenium. As we have mentioned before in Chap. 1, after the Second World War, germanium and silicon crystals became the center of attention, in particular due to the research effort at the American Bell Laboratories. Both substances consist of only a single element. Their crystalline structure is the same as that for diamond, being much simpler than that for copper-oxide and selenium. As a new chemical element, germanium had been discovered in 1886 by the German chemist Clemens Winkler working at the Mining Academy in the Saxonian town of Freiberg. Germanium and silicon are located in the fourth group of the Periodic Table with carbon at the top and silicon directly below. Carbon and silicon are among the most abundant elements on the earth. Directly below silicon in the Periodic Table one finds germanium, which is much more rare and which until its discovery only existed in the form of an unoccupied spot in the Periodic Table.

At room temperature the concentration of the thermally activated charge carriers in the conduction band and in the valence band of germanium is about one billion times smaller than in a good electrical conductor such as, say, copper. However, not far above room temperature so many additional charge carriers are thermally activated in germanium, that its electronic properties change too much for many electronic applications. Even hot summer temperatures can no longer be tolerated. In silicon the energy gap between the valence band and the conduction band is nearly twice as large as in germanium. Therefore, the corresponding concentration of the thermally activated charge carriers in silicon is about ten thousand times smaller than in germanium. Hence, silicon reacts much less sensitively to summer temperatures. Because of this fact and in particular also because of the spontaneous growth of a thin and stable layer of silicon-oxide, acting as an excellent electrical insulator on its surface, silicon is far superior as semiconductor material for most electronic applications compared with germanium and today dominates the semiconductor industry.

> In Chap. 1 we mentioned the invention of the transistor by John Bardeen, Walter Brattain, and William Shockley in the year 1947. At this time a completely new and unknown territory had been entered, and many important new experiences had to be gained for the first time. Also at the beginning, the extreme requirements regarding the purity of the semiconductor materials were by no means clear. The recognition that chemical impurities and grain boundaries in the crystals strongly affect the electrical properties was only gradually accepted. It was Gordon Kidd Teal in the American Bell Laboratories who was convinced very early on, that only a large effort in the preparation and purification of single crystals could lead to success. However, at first nobody wanted to listen to him. Therefore, only as an outsider and after many difficulties was he able to push forward his ideas for growing ultra-pure single crystals. Today the fabrication of large, ultra-pure single crystals of silicon as the raw material for the semiconductor industry is performed worldwide on a large technical scale and represents an important business (for example, by the Wacker Chemie AG in Burghauen, Bavaria; Fig. 1.6). In the early days Gordon Teal was soon hired away from the Bell Laboratories, and since January 1953 he pursued his ideas in the American company Texas Instruments. This Company then developed into the largest semiconductor manufacturer of the world.

In addition to semiconductors such as germanium and silicon which consist only of a single element from the fourth group of the Periodic Table, there also exist other substances which are composed from several elements and which are highly interesting for technical applications because of their semiconductor properties. The pioneering ideas about these "compound semiconductors" were developed by the German Heinrich Welker, in the early 1950s. At the time Welker worked at the Siemens Research Laboratory in Erlangen. Later he became the director of this Laboratory. Previously, Welker had been an assistant of Arnold Sommerfeld at the University of Munich, where he had worked among other things on the theory of superconductivity. In Erlangen he wanted to develop a better understanding of the semiconductor physics of germanium, and within this context he considered the following question. In the germanium atom there are four electrons in the outer shell. Is it perhaps possible, that a compound of an atom with five electrons from the fifth group of the Periodic Table and an atom with three electrons from the third group will also show semiconductor properties very similar to those of germanium, since on the average we have again four electrons per atom as in germanium? On the left side of germanium in the Periodic Table we find gallium and on the right side, arsenic. The experiments then confirmed indeed, that the compound gallium-arsenide (GaAs) is a semiconductor. Indium-antimonide (InSb) is similarly another member of the group of the "III-V semiconductors". The III-V semiconductors displayed interesting electronic properties. The mobility of the electrons and the holes was much larger than in germanium and silicon. Therefore, technical applications became possible which required a faster response of the charge carriers. Furthermore, in the III-V semiconductors, the energy gap between the valence band and the conduction band is relatively large. This is highly interesting for applications in optoelectronics, as will be discussed further below. The principle of the compound semiconductors has subsequently also been extended to compounds of elements from the second and the sixth group of the Periodic Table. The latter compounds are referred to as "II-VI semiconductors".

Very recently, peculiar band-structure effects and contact points between the valence band and the conduction band have attracted much theoretical and experimental attention. We will come back to these developments in Sect. 11.6, when we turn to topological insulators, metals, and semimetals.

6.2 Doped Semiconductors

So far we have restricted our discussion to semiconductors consisting of only a single element or being a compound of exactly two elements. Other substances acting as additives have been excluded. In this way we were dealing only with the case of intrinsic semiconductors. However, the case where a semiconductor is doped with foreign atoms is much more important. Next we turn to this case of the "extrinsic semiconductors". It was noted already in the early days that the electrical properties of nominally the same semiconductor material fluctuated within wide limits, such that an exact reproducibility was impossible. Because of the extremely low concentration of mobile charge carriers in the intrinsic semiconductors compared with metals, the electrical properties of the former are extremely sensitive against chemical impurities or defects in the crystal lattice. As we have discussed before, silicon and germanium belong to the fourth group of the Periodic Table and, hence, possess four electrons in their outer atomic shell. However, if we incorporate atoms from the fifth group of the Periodic Table, having five electrons in the outer shell, as, for example, phosphorus or arsenic into the silicon or the germanium lattice, the fifth electron represents a surplus. This excess electron can easily be transferred by means of thermal excitation from the phosphorus or the arsenic atom into the conduction band of silicon or germanium. The phosphorus or the arsenic atom then remains in the host lattice of silicon or germanium as a singly and positively ionized atom. Because of the donation of their excess electrons to the conduction band of the host lattice, these incorporated atoms from the fifth group are denoted as "donors". Doping of the semiconductor with these donors allows the concentration of the charge carriers in the conduction band to be changed by many orders of magnitude, compared with the intrinsic semiconductors such as silicon and germanium (Fig. 6.2).

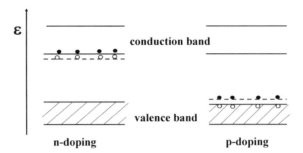

Fig. 6.2 Doped semiconductors. The conduction band is separated from the valence band by a relatively large energy gap. With n-doping, electrons are thermally excited from the donors up to the lower edge of the conduction band. With p-doping, the thermal excitation of the electrons occurs from the upper edge of the valence band up to the energy levels of the acceptors, such that holes remain in the valence band

The basic idea underlying the concept of donors can be extended further. Therefore, next we consider the doping of the host lattice of silicon or germanium with atoms from the third group of the Periodic Table, having only three electrons in their outer shell, such as, for example, aluminum or gallium. Now the incorporated atom possesses one electron less than the atoms of the host lattice. The missing fourth electron can be captured by the incorporated atom by means of thermal excitation from the valence band of silicon or germanium. At the same time a hole appears near the upper edge of the valence band of the latter. The aluminum or gallium atom then remains in the host lattice of silicon or germanium as a singly and negatively ionized atom. Because of this acceptance of their missing fourth electrons from the valence band of the host lattice, these incorporated atoms from the third group are denoted as "acceptors". As we have discussed above, the holes near the upper edge of the valence band also participate in the electrical conduction mechanism in the same way as the electrons in the conduction band. Again, the doping of the semiconductor with the acceptors allows one to change, in a controlled way, the concentration of the mobile holes, acting as charge carriers in the valence band, by many orders of magnitude compared with the intrinsic semiconductors (Fig. 6.2).

These concepts of the donors and acceptors were already developed in the early 1930s, and they are still valid today. Important early contributions came from Rudolf E. Peierls and Allan H. Wilson, mentioned before in Chap. 4. At the time, the German theoretical physicist Walter Schottky also helped to clarify the underlying physics. It is possible to vary the concentration of the mobile charge carriers over many orders of magnitude by means of doping, which makes the semiconductors so interesting for electronic applications. In contrast to the intrinsic semiconductors, the doped semiconductors are referred to as extrinsic semiconductors. Furthermore, the extrinsic semiconductors are denoted according to their kind of doping: Semiconductors doped with (negative) electrons from the donors are referred to as n-doped, and those doped with (positive) holes from the acceptors are referred to as p-doped.

6.3   Excitons and Electron-Hole Droplets

The population of the conduction band at its lower edge with electrons and of the valence band at its upper edge with holes cannot only be accomplished by means of the thermal excitation of electrons. This can also be achieved under light irradiation due to the absorption of light quanta. Because of the irradiation with light, the electrical conductivity can be significantly increased, a phenomenon referred to as photoconductivity. This effect is technically utilized in light meters in the form of photo cells. This optical excitation is particularly interesting in intrinsic semiconductors. If an electron is energetically raised from the valence band into the conduction band due to the absorption of a light quantum, an electron-hole pair is generated. Both particles possess opposite electric charges and can form a bound state, similar to the electron and proton of the hydrogen atom. In the bound state both particles move around their common center of gravity. This bound configuration of an electron-hole pair is denoted as an exciton. The excitons can move around within the crystal and transport excitation energy in doing so. However, they do not transport electric charge, since they are electrically neutral, their total charge being zero. By recombination, both particles of the exciton can annihilate each other. The energy which is set free during this process mostly appears again in the form of an emitted light quantum.

The generation of excitons under light irradiation is strongly enhanced if the crystal is cooled to low temperatures. Within the crystal the excitons behave like a gas, which condenses and forms a liquid at sufficiently low temperature and sufficiently high density. The electron-hole droplets or the electron-hole liquid have been studied, in particular in germanium, at low temperatures where the traces of the emitted light have been utilized in impressive experiments (Fig. 6.3).

Fig. 6.3 Energetic excitation of an electron from the upper edge of the valence band into the conduction band by means of the absorption of a photon. **a** An electron-hole pair generated in this way can assume the bound state of an "exciton". **b** In a semiconductor crystal, at low temperatures and sufficiently high concentration, the excitons condense into droplets of the electron-hole liquid

6.4   Metal-Semiconductor Contact, P-N Junction

The fact, that the concentration of mobile charge carriers in semiconductors is smaller by many orders of magnitude compared with metals, leads to novel phenomena and to the dependence of the electric current flow upon the current direction already observed by Ferdinand Braun. At the location of the junction between a metal wire and the semiconductor crystal there occurs a depletion of the mobile electrons and holes, and an electrically insulating boundary layer is generated in the semiconductor. The underlying theoretical model concepts were developed by Walter Schottky, employed by the Siemens Company in Germany (Fig. 6.4). Therefore, one also speaks of the Schottky diode and of the Schottky boundary layer.

In order to establish equilibrium between both sides of the contact, electrons flow from the semiconductor (we assume an n-doped semiconductor) into the metal. As a result, there develops a positive space charge, which extends in the semiconductor over a finite distance. At the junction, this leads to an increase of the potential, which must be overcome by the electrons during the flow of electric current, and which depends on the voltage V applied to the contact. Hence, the contact acts as a rectifier, where the current I is given by

$$I = I_S(e^{eV/k_B T} - 1) \tag{6.13}$$

Here I_S is the saturation current.

The electric current can flow across this metal-semiconductor contact only when, for an n-doped semiconductor, the free electrons, or for a p-doped semiconductor,

n-doped semiconductor   **metal**

Fig. 6.4 Walter Schottky (*Photo* Deutsches Museum). Metal-semiconductor contact, also called Schottky contact (right). At the junction between an n-doped semiconductor and a metal, within the semiconductor a depletion zone of the mobile electrons is generated, in which the positively ionized donors remain behind. At the adjoining metal surface, negative charges accumulate. In the case of a p-doped semiconductor, negatively ionized acceptors remain behind in the depletion zone the free holes, are moving from the semiconductor into its depletion zone, thus filling up the depletion zone. In the opposite current direction the electrically insulating boundary layer remains unchanged, and the current flow is interrupted. In this way the rectifying effect of the metal-semiconductor contact is accomplished. Schottky's first paper on this subject appeared in 1923. He published his complete theory about the barrier layer and the point-contact rectifier during the years 1939–1942, partly in collaboration with Eberhard Spenke, who was also working for the Siemens Company. Because of the rectifying properties of the metal-semiconductor contact we have discussed, special procedures are necessary for supplying semiconductor circuits with electric current. For this purpose ohmic contacts consisting of heavily doped semiconductor regions, referred to as n^+ or p^+ regions, proved to be quite satisfactory.

In addition to the metal-semiconductor contact, the junction between an n-doped and a p-doped semiconductor, the p-n junction, had also received much attention (Fig. 6.5). On the n-doped side of the junction there exist many electrons in the conduction band, whereas on the p-doped side many holes populate the valence band. However, the large difference in concentration of the particular charge carriers, respectively, between both sides must be balanced, since at equilibrium the strong concentration gradient of the electrons and of the holes down to the opposite side of the junction cannot be maintained. Therefore, the electrons diffuse from the n-doped into the p-doped region, and the holes diffuse in the opposite direction. As a result, in the n-doped region, positively ionized donors and in the p-doped region negatively ionized acceptors, remain in the form of space charges. In this way a local electric field is generated, exercising a force on the charge carriers in the

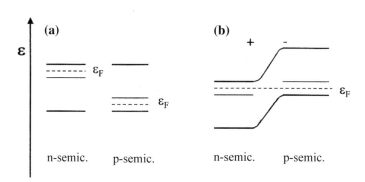

Fig. 6.5 p-n junction. **a** As long as there exists no connection between the n-doped and the p-doped semiconductor, the energy diagram displays clearly different values of the Fermi energy ε_F in the two semiconductors. **b** With an electric contact between the n-doped and the p-doped semiconductor, the concentration gradient of the electrons and of the holes between both sides of the junction is equalized by means of diffusion of both kinds of charge carriers to the opposite side, respectively. During this process, positive or negative space-charge regions are generated on both sides of the junction, resulting in a local electric field between both sides. The diffusion process ends when the Fermi energy ε_F has reached the same value on both sides of the junction

opposite direction to the driving force of the two diffusion processes, respectively. Eventually, because of the local electric field, the diffusion processes come to a complete stop. However, at the location of the junction there remains now an electrical potential gradient. Depending upon the direction of the electric current, the potential gradient at the p-n junction increases or decreases because of the current. Therefore, a rectification effect is achieved again, similar to the metal-semiconductor contact.

At the boundary of the p-n junction there develops a *dynamic equilibrium* between two processes acting in opposite direction. (The thickness of the boundary layer is typically 1 μm, being much smaller than the diffusion length of the charge carriers.)

1. Diffusion of electrons (holes) into the p-region (n-region), where they disappear because of recombination. This generates the *recombination current*

$$I_r = I_{nr} + I_{pr} \tag{6.14}$$

2. Thermal excitation of electrons (holes) in the p-region (n-region), which fall down into the n-region (p-region) on the other side of the junction and cause the thermal *generation current*

$$I_g = I_{ng} + I_{pg} \tag{6.15}$$

(We note that in our energy scheme the energy of the positive charges decreases in upward direction.) In thermodynamic equilibrium (zero applied voltage) we have

$$I_{nro} + I_{ngo} = 0; \quad I_{pro} + I_{pgo} = 0; \quad I_{ro} + I_{go} = 0 \tag{6.16}$$

Now we turn to the rectification effect of a p-n junction. We assume that nearly all of the applied voltage V appears across the boundary layer, since the total carrier concentration n + p is a minimum there, and the electric resistance is a maximum. We start with the *reverse voltage bias* (Fig. 6.6a). Now the n-region (p-region) is on the positive (negative) side, and in the p-region the potential is raised further. This additional potential eV causes a reduction of the recombination current by the Boltzmann factor:

$$I_{nr} = I_{nro} \exp(-eV/k_B T); \quad I_{pr} = I_{pro} \exp(-eV/k_B T) \tag{6.17}$$

The thermally excited generation current is not affected by the raising or lowering of the band along the energy axis:

$$I_{ng} = I_{ngo}; \quad I_{pg} = I_{pgo} \tag{6.18}$$

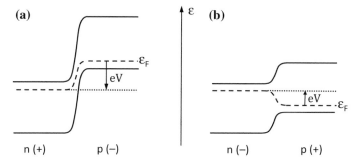

Fig. 6.6 Rectification effect of a p-n junction. **a** Reverse voltage bias. **b** Forward voltage bias

The total current in reverse direction is

$$I = I_{nr} + I_{ng} + I_{pr} + I_{pg} \tag{6.19}$$

With (6.16) and (6.18) we obtain

$$I = (I_{nro} + I_{pro})[\exp(-eV/k_BT) - 1], \text{ and} \tag{6.20a}$$

$$I = I_S[\exp(-eV/k_BT) - 1] \tag{6.20b}$$

where $I_S = I_{nro} + I_{pro}$ is the saturation current.

In the case of a *forward voltage bias* (Fig. 6.6b) with the n-region (p-region) being negative (positive), the potential in the p-region is *lowered* by the amount eV. Now the recombination current is *increased* by the Boltzmann factor (with negative energy in the exponent):

$$I_{nr} = I_{nro} \exp(eV/k_BT); \quad I_{pr} = I_{pro} \exp(eV/k_BT) \tag{6.21}$$

Again, the generation current remains unaffected. Analogous to the reverse bias, we obtain

$$I = I_S[\exp(eV/k_BT) - 1] \tag{6.22}$$

In Fig. 6.7 we show the typical rectifier characteristic expected from (6.20a, 6.20b) and (6.22). We note that in forward direction I/I_S increases with decreasing temperature.

The American Russell Shoemaker Ohl working at the Bell Radio Laboratories in Holmdel in the Federal State of New Jersey had already been concerned in the 1940s with p-n junctions in silicon. He was interested in the application of these junctions as possible crystal detectors for radio- and microwaves. Incidentally, he also discovered their interesting photovoltaic properties, as we will discuss below. At the time in many companies within the electronics industry people started to investigate the rectifying properties of p-n junctions. For example, soon after the

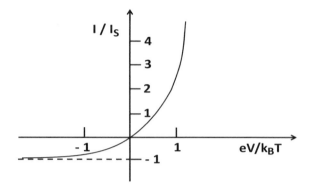

Fig. 6.7 Rectifier characteristic of a p-n junction: current I plotted versus voltage V in reduced units

Second World War the Siemens Company started a laboratory for this purpose in the small town of Pretzfeld near Erlangen in Frankonia. Under the management by Eberhard Spenke rectifiers based on selenium were investigated at first and then fabricated in a pilot plant. Eventually, the optimizing process resulted in a contact between p-doped selenium and n-doped cadmium-selenide (CdSe). The old-fashioned rectifiers, operating with mercury vapor and installed for the high-current applications of power electronics, were now replaced by the selenium rectifiers. Furthermore, there existed a multitude of technical applications of the selenium rectifiers in the low-current technology in the field of radio and of electronic communication. During 1952 these developments in Pretzfeld, based on selenium, were stopped, since germanium and silicon moved to the forefront. Subsequently, rectifier development for power electronics concentrated on germanium and silicon.

6.5   Transistor

P-n junctions are also the basis of the "bipolar transistor". A transistor operates like a valve, by which the electric current flow is electronically controlled from the outside. Hence, it has three connections to the outside: input, control, and output. Originally, in their invention John Bardeen and Walter Brattain had used an arrangement, which later became known as point-contact transistor (Fig. 6.8a). Two gold contacts were pressed upon an n-doped germanium crystal within a mutual distance of only 50 μm. A third metal contact, the "base", was attached to the back of the germanium crystal. One gold contact acted as the "emitter" and served for injecting holes into the germanium crystal. The other gold contact collected the holes again and is referred to as the "collector". The electric current flow between the emitter and the collector can be modulated by changing the electric potential at the emitter versus the potential at the base and at the collector. The first experiments with this arrangement had already yielded a current amplification of about 40 and a voltage amplification of about 100. This success clearly opened the way to replace

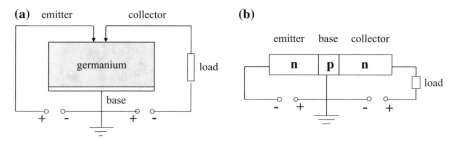

Fig. 6.8 Transistor principle: **a** Point-contact transistor as it was used originally by Bardeen and Brattain. Two gold contacts are pressed upon an n-doped germanium crystal within a distance from each other of only about 50 μm. On the opposite side of the germanium crystal a third metal contact, the "base", is attached. One gold contact acts as emitter and injects holes into the germanium crystal. The holes are collected again by the other gold contact, acting as the collector. **b** Junction transistor according to Shockley. A p-doped semiconductor is placed between two n-doped semiconductors in such a way, that two p-n junctions are formed mirror-symmetrically. Whereas one n-doped semiconductor serves as the emitter, the other n-doped semiconductor acts as the collector. The p-doped region in the center functions as the base

the evacuated glass tube by a solid-state device to be used for electronic amplification (Fig. 6.9).

Soon after the first demonstration of the transistor principle based on the point-contact transistor William Shockley proposed another version of the bipolar transistor, which is based on two p-n junctions. One p-n junction operates as an emitter, whereas the other p-n junction is electrically connected in the opposite direction and serves as the collector. This "junction transistor" of Shockley consists of three regions: an n-doped or a p-doped central region, which is acting as the base and which on both sides is joined to a region with the opposite doping, respectively. In this way a p-n junction is formed on both sides of the central region (Fig. 6.8b). The operating principle is similar to that of the bipolar point-contact transistor.

Again, "minority charge carriers", having the opposite charge to that corresponding to the doping of the particular region, are injected by the emitter into the central region and are then taken up again by the collector. The current of these minority charge carriers can be modulated again by means of changes in the electric potential. In the junction transistor of Shockley all crucial semiconductor functions are now transferred from the surface into the interior of the crystal. The highly sensitive crystal surface no longer has the central role. Since in the transistor operation the negative electrons as well as the positive holes are utilized, one refers to bipolar transistors.

In addition to the two types of transistor from the early days which we have briefly discussed, subsequently many more versions were proposed and studied experimentally. In the meantime the transistor has gone through many stages of evolution. During this development its rapidly progressing miniaturization has always been a strong driving force. Furthermore, the transistor had to operate faster and faster, allowing its use at higher and higher frequencies. As an electronic

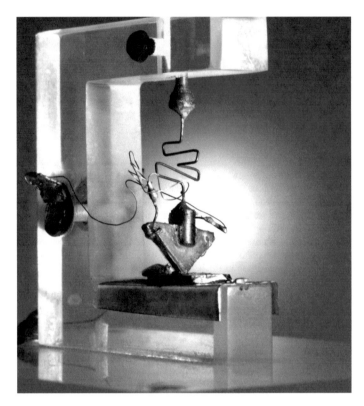

Fig. 6.9 The first point-contact transistor constructed by Bardeen and Brattain in December of 1947. The three-cornered part in the middle is made from plastic material and is covered by a gold foil at its two edges. At the tip at the bottom the gold foil is cut by a razor blade, such that two contacts are generated in close proximity. By means of a metal spring the contacts are pressed against the semiconductor surface located underneath (*Photo* Lucent)

device, the transistor has completely replaced the evacuated amplifier tube made from glass. Compared with this forerunner of glass, the transfer of the electronic functions into the crystal interior achieved by the transistor, yielded important advantages: highly increased reliability and robustness, as well as the potential for extreme miniaturization and, hence, for fabrication in large quantities and at a very low price.

If we look at the commercial use of the invention of the transistor, we can observe a very interesting process. Initially, the company, which owned the invention, pursued the following guiding principle: keep it secret and do not divulge any details. However, after some time the management noticed that applications of the transistor did not appear, and that scepticism still dominated. It became very clear, that the strategy of the company had to be changed. Therefore, a complete turnaround was adopted, and the attitude now became quite open. As a result, during September 1951 the Bell Laboratories organized a large symposium in Murray Hill, in which the details of transistor technology were openly discussed. The event was met with strong interest, and 301 professional people participated. The participants came from universities, from other industrial laboratories, and from military organizations. The Proceedings Volume with the Conference Reports contained 792 pages, and 5500 copies were distributed. This great success and the rapidly increasing general interest were the reason why, during April 1952, a second symposium on transistor technology was organized. This time representatives from a total of 40 companies participated: 26 companies from the USA, and 14 companies from foreign countries. Without doubt, it was this policy of the Bell Laboratories of switching to an open attitude which began the decisive change. Now ideas and proposals came from many sides, including from outsiders. The first commercial application of the transistor was in hearing aids made by American companies. Subsequently, the technical utilization of the transistor has grown rapidly.

Already in 1926, the Jewish Austro-Hungarian-born physicist Julius Edgar Lilienfeld had patented a field-effect transistor, which kept the Bell Laboratories from obtaining a patent for its discovery.

6.6   Photovoltaics, LED, Semiconductor Laser

The American Russel S. Ohl had discovered the photovoltaic effect of the p-n junction rather by accident (Fig. 6.10). If the p-n junction is irradiated with light, an electric voltage appears between both sides, or an electric current flows through a wire connecting both sides around the outside. This is exactly the principle of the solar cell. The absorption of a light quantum causes the generation of an electron-hole pair at the p-n junction. Because of the electrical potential gradient at the p-n junction, the electrons are accelerated towards the n-doped region and the holes towards the p-doped region. As a result, an electric current, flowing through the wire around the outside, is generated. For the large-scale technical utilization of the solar energy the search for increasing efficiency of the solar cell today is still an

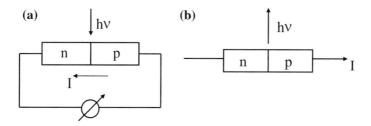

Fig. 6.10 Photovoltaic effect at the p-n junction. **a** Solar cell: The absorption of photons of sufficiently high energy within the region of a p-n junction results in the generation of electron-hole pairs. In the electric potential drop of the p-n junction, the electrons and the holes lead to electric current flow through the conductor and around the outside. **b** Inversion of the solar cell in the light emitting diode (LED). An electric current flowing in the forward direction injects electrons into the p-doped semiconductor and holes into the n-doped semiconductor. By means of the recombination of the electrons with the holes energy is released, which is emitted in the form of photons

important subject of research and development. For transportation in space, the silicon solar cell represents the major energy source today.

The process we have just discussed can also be inverted. Then we are dealing with the light- emitting diode (in short LED), and an additional step in the development leads us to the injection laser or the semiconductor laser. Now an electric current is passed through the p-n junction in a forward direction. As a result, electrons are injected into the p-doped region and holes into the n-doped region. Being minority charge carriers, the electrons recombine with the holes in the p-doped region, and the holes do the same with the electrons in the n-doped region. The energy, which is set free during the recombination, is emitted in the form of a light quantum. In this way the light emitting diode is accomplished. However, for the semiconductor laser additional requirements must be fulfilled. As is well known, light emission always occurs by means of a quantum mechanical process, in which an electron as an atomic elementary particle drops from a higher to a lower energy level. However, for the generation of laser light it is necessary that the upper energy level is occupied by more electrons than the lower level. We must have "population inversion". As an additional requirement, a standing light wave must be built up in the active region of the p-n junction in the form of resonance due to the proper geometric dimensions. Finally, all possible competing processes, not resulting in the emission of a light quantum and following a different course during the electron-hole recombination, must be sufficiently suppressed.

The operation of the first semiconductor laser was in 1962. This new technology in the field of optoelectronics benefitted greatly from compound semiconductors, which Heinrich Welker had discovered about a dozen years earlier, and which are most suitable for this application. In the III-V semiconductors, as for example gallium-arsenide (GaAs), the energy gap between the conduction band and the valence band is relatively large. Hence, the energy being set free during the electron-hole recombination in the form of a light quantum is also correspondingly large. The spectrum of the visible light extends from red on the end for the light quanta with low energy, up to blue and violet on the other end for the light quanta with high energy. Gallium-arsenide emits light in the invisible infrared. Red and green light is generated by mixed crystals based on gallium-arsenide containing further admixtures. Only recently, it created a small sensation when Shuji Nakamura in Japan succeeded for the first time in building a semiconductor laser based on gallium-nitride (GaN) emitting even blue light. Meanwhile semiconductor lasers such as the GaAs laser have found wide application in many areas. Using the infrared laser made from gallium-arsenide, we operate the remote control of our television receiver. In many household items we find red and green little lights fabricated from semiconductor crystals. Over the years, the yield of the emitted light could be strongly improved by means of a modification of the semiconductor material on both sides of the active p-n junction region, achieving an energetic spatial confinement of the electrons and of the holes within a small, well defined active volume. In this case one speaks of the double hetero junction (in short DH).

As early as in 1963 the German Herbert Kroemer had proposed this advanced type of semiconductor laser. However, at that time it had not yet been recognized that the optoelectronics eventually would gain that much in importance. Therefore, Kroemer's ideas were ignored for quite a while. The remarks Kroemer made in December 2000 in Stockholm during his lecture celebrating his award of the Nobel Prize are extremely noteworthy. In this lecture Kroemer said:

> .... It was really a classical case of judging a fundamentally new technology, not by what new applications it might *create*, but merely by what it might do for already existing applications. This is extraordinarily shortsighted, but the problem is pervasive, as old as technology itself. The DH laser was simply another example in a long chain of similar examples. Nor will it be the last. .... Any detailed look at history provides staggering evidence for what I have called .... the *Lemma of New Technology*: The principal applications of any sufficiently new and innovative technology always have been – and will continue to be – applications *created* by that technology.

Very recently, light bulbs are less and less used for general illumination purposes and are replaced by "cold light sources". In the case of these energy saving illuminants, the light emitting diodes (LED) play an important role (Fig. 6.10b). Whereas their lifetime is relatively long, their production costs are still high. Compared with a traditional light bulb, the energy saving of the LEDs at best amounts to 89%. Presently, the optimum color composition of the light is a subject of ongoing research.

6.7   Miniaturization, Planar Technology

During the past 60 years, semiconductor electronics has reached an impressively high level and has become an important economic field. Initially, the methods for the preparation of the semiconductor materials looked more like black art, consisting of many special tricks and recipes. However, going through many evolutionary stages, eventually they developed into the industrial processes used today in semiconductor factories. This development was accompanied by the permanently progressing miniaturization, which allowed the placement of an ever increasing number of devices and electronic circuits within an area of about 1 cm^2 on a chip. An important advance has been the introduction of "planar technology", utilizing the silicon surface well protected by its highly stable oxide. A large single crystal of silicon is cut into thin slices, the "wafers". The thickness of the wafers is only a few tenth of a mm. All further processing steps are concerned only with the silicon surface. The doping with donors and acceptors is accomplished by means of diffusion into the regions near the silicon surface, where the oxide had been etched away previously. For this purpose the temperature of the diffusion ovens must be exactly controlled within a fraction of one degree, in addition to the temporal profile of the heat treatment. Only with such extreme care can the doping profiles near the surface of the wafer be exactly reproduced. The many processing steps, which today can amount up to more than four hundred, are controlled by computers. Today, the diameter of the silicon wafers and, hence, the diameter of the silicon single crystals used as the raw material has reached a value as large as 30 cm

Fig. 6.11 Processing of the silicon wafers with 30 cm diameter (*Photo* Wacker Chemie AG)

(Fig. 6.11). From a single wafer with this diameter a total of about 700 chips each with 1 cm^2 area can be fabricated. At the end of the fabrication process, in some cases a single wafer of this kind can reach a total value of up to 250 000.- US $ (Fig. 6.12).

Due to the permanently progressing miniaturization, the number of transistors on a single chip has increased rapidly. During the three decades from 1970 until 2000 this number increased from about one thousand at the beginning up to about 256 million at the end. The cost per bit of stored information dropped correspondingly. This highly impressive development has been summarized in the famous law of the American Gordon E. Moore, about which he had been contemplating already in 1965. According to this law, every 5–6 years the price per transistor on a chip drops to about one tenth of its value at the beginning of this period. However, the increasing complexity of the semiconductor circuits is accompanied by a corresponding increase in the cost of the semiconductor factories. Sometimes, this fact is also referred to as the Second Moore's Law.

Today (2012), the individual "structure size" for the geometric dimensions of the devices to be fabricated is reaching about 45 nm. Before long, it is expected to hit a principal limit because of the atomic or molecular quantum conditions. In order to extend the miniaturization any further, completely new concepts will then become necessary.

During recent years, an unexpected and completely different development took place, which started from the technical experience with silicon single crystals, and which turned out to be highly interesting. It has nothing to do with the electronics of doped semiconductors. Instead, it is concerned with the technical utilization of single-crystalline silicon in the field referred to as micromechanics. By now, micromechanics has turned into an important new technical field experiencing rapid

Fig. 6.12 Modern microprocessor chip with the dimensions 12.6 mm × 12.6 mm. (Here the microprocessor Power 3). There are about 15 million transistors on this chip. (IBM)

development. Extremely miniaturized mechanical systems are fabricated by means of different etching techniques and other methods of micro-fabrication. For example, such systems are used for measuring pressure or mechanical acceleration. Actual applications of these systems can be found, among other areas, in the automobile industry. Here, a typical example is the controlled activation of the airbag. Again, also in this case, the potential for fabrication of large quantities together with a very low price represents an important requirement.

6.8   Thermoelectricity, Peltier Cooling

At the end of this chapter dealing with the properties of semiconductors we turn now to thermoelectric phenomena, namely the Peltier and Seebeck effect. As we have discussed in Chap. 5, both effects appear if a temperature gradient and an electric potential gradient act simultaneously. In semiconductors both effects are much stronger than in metals, typically by a factor of one hundred or more. This fact is due to the particular form of the Fermi distribution of the electric charge carriers in semiconductors. As we have pointed out above, in semiconductors the number of mobile charge carriers is much smaller than in metals. Therefore, the Fermi energy ε_F is also correspondingly smaller in semiconductors. As we have also discussed in Chap. 5, the states are always occupied by electrons according to the Fermi function. The energy width $k_B T$, within which the Fermi function drops from the value of one to zero, in semiconductors is much larger than ε_F. This is in contrast to metals, where this energy width is much smaller than ε_F. Therefore, in semiconductors the Fermi distribution (5.10) changes into the classical Boltzmann distribution (6.1). As a consequence, the reduction factor $k_B T / \varepsilon_F$, which is valid for metals and which selects only a small fraction of all electrons to participate in many thermal and electrical phenomena, disappears in semiconductors. This is the main reason for the relatively high values of the Peltier and Seebeck effect in semicondutors.

In particular, the Peltier effect in semiconductors is very appropriate for technical applications. We recall that the Peltier effect is due to the transport of the heat energy of the charge carriers moving through the conductor during electric current flow. At the junction between two different electrically conducting materials a pile-up of the heat current can develop. Depending upon the current direction, the junction region can be heating up or cooling down. This effect is most pronounced, if an n-doped and a p-doped semiconductor join together at the junction. In this case the (negative) electrons of the n-doped side and the (positive) holes of the p-doped side move in opposite directions. Hence, they move either from both sides towards the junction, or away from the junction on both sides. In the second case a strong effective cooling of the junction is expected. Therefore, the Peltier effect is useful in cryogenics.

The Russian Abram Fedorovich Ioffe has been one of the first who recognized the importance of semiconductors in cryogenics. Ioffe was born in the county town of Romny, at the time belonging to the Russian Empire. During the years 1902 – 1906 he was probationer at first and assistant later of Wilhelm Röntgen at the University in Munich. In 1905 Ioffe finished his Ph.D. thesis, supervised by Röntgen, entitled "Elastic After-effect in Crystalline Quartz". During the Fall of 1906 Ioffe took the position as assistant at the Polytechnic Institute in St. Petersburg. This Institute, named after Peter the Great, had been founded in 1902. During his exceptional scientific career Ioffe was responsible for the establishment of five different Research Institutes in the former Soviet Union. Here we want to emphasize in particular the Semiconductor Institute in St. Petersburg, which became very famous, and from which many important papers on the physics of semiconductors originated. Since 1950 Ioffe strongly increased his research effort in the field of thermoelectricity in semiconductors, then still at the Physical-Technical Institute of Leningrad. A few years later he wrote a book on semiconductor thermo-elements. At the time, his optimistic forecast about the thermoelectric applications of semiconductors, in particular for cryogenics due to "Peltier cooling", has triggered great new efforts in semiconductor research worldwide in many laboratories of the electronics industry.

Today, Peltier cooling is carried out primarily based on the n-doped and the p-doped semiconductor compound bismuth-telluride, Bi_2Te_3. The commercially available "Peltier modules" consist of an arrangement of up to several hundred n-doped and p-doped Peltier legs, thermally in parallel and electrically in series connection. The individual Peltier legs are only a few mm long and have a cross-section of about 1 mm². With a single Peltier module, a cooling from room temperature down to 50–60° below room temperature can be achieved. Even lower temperatures can be reached using a multi-stage Peltier cascade (Fig. 6.13). For example, recently, cooling from room temperature down to 135 K was accomplished by means of a seven-stage Peltier cascade.

The operation of a stage for Peltier cooling, schematically shown on the left in Fig. 6.13, is governed by the power-balance equation

$$|S|\, T_1\, j = \kappa \frac{T_0 - T_1}{L} + \frac{1}{2} \frac{V^2}{\rho L} \qquad (6.23)$$

which describes the heat current density flowing into or out of the cold end of the Peltier element. On the left in (6.23) we have the density of the Peltier heat current. The first term on the right is the heat current density caused by thermal conduction. The second term on the right represents the dissipated electrical power per cross-sectional area, half of which is assumed to reach the cold side of the Peltier element. Equation (6.23) applies to each of the two legs of the Peltier element. T_0 and T_1 are the temperature on the warm and the cold side, respectively. ($|S|$ = absolute value of the Seebeck coefficient, κ = heat conductivity, ρ = electric resistivity, V = voltage, L = length of the Peltier legs. These quantities are assumed to be the same in both legs. $j = (1/\rho)\,(V/L)$ = electric current density). In (6.23) any additional thermal load of the cold end is assumed to be zero. From (6.23) one finds

$$(T_0 - T_1) = \frac{1}{\kappa \rho}\left[|S|\, T_1\, V - \frac{V^2}{2}\right] \qquad (6.24)$$

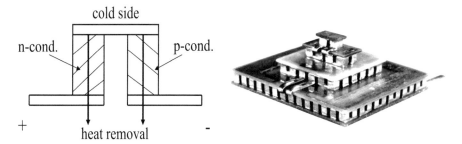

Fig. 6.13 Peltier cooling. Left: Schematics of a Peltier cooling device consisting of an n-doped and a p-doped semiconductor. The electric current flows from the left to the right side. Right: Commercially available four-stage Peltier cascade. In the individual stages the Peltier legs are clearly visible. With the cascade shown, the temperature can be lowered from room temperature at the warm end down by about 140 K lower than room temperature at the cold end. The lateral dimensions are: lowest stage: 24.0 mm × 20.6 mm; uppermost stage: 4.5 mm × 2.4 mm. Total height: 13.6 mm (*Photo* KRYOTHERM)

The maximum of $(T_0 - T_1)$ is obtained from the condition $\partial(T_0 - T_1)/\partial V = 0$, yielding $V = |S|\, T_1$ and finally

$$(T_0 - T_1)_{max} = \frac{1}{2} z\, T_1^2 \qquad (6.25)$$

Here

$$z = \frac{S^2}{\kappa\, \rho} \qquad (6.26)$$

is the "figure of merit", a most important quantity of thermoelectricity.

As we see from (6.24), it is essentially the combination of the linear current dependence of the Peltier heat and the quadratic current dependence of the generated Joule heat, which limits the cooling effect achieved by the Peltier element, and which leads to a maximum temperature drop. If the cold side experiences an additional heat load (as, for example, from the higher stages of a Peltier cascade), then the maximum temperature drop will be smaller than the value given in (6.25). With increasing such heat load, the temperature drop $(T_0 - T_1)$ decreases linearly, reaching zero when the heat load is equal to half of the electric power dissipated in the Peltier stage.

Equation (6.25) can be rewritten in the form

$$(T_0 - T_1)_{max} = \frac{1}{2z} \left[(2\, z\, T_0 + 1)^{1/2} - 1 \right]^2 \qquad (6.27)$$

Using the typical values near room temperature (applicable to the Bi_2Te_3 system mentioned above), $S_n = -200\ \mu V/K$, $S_p = 200\ \mu V/K$, $\rho = 1\ m\Omega\ cm$, and $\kappa = 15 \cdot 10^{-3}\ W/cm\ K$, one obtains $z = 2.7 \cdot 10^{-3}\ K^{-1}$. This value of z together with (6.27) and $T_0 = 283\ K$ yields $(T_0 - T_1)_{max} = 64\ K$.

Chapter 7
Circling Electrons in High Magnetic Fields

Abstract In a magnetic field the motion of electric charges perpendicular to the field direction is deflected into circular orbits because of the Lorentz force. This deflection causes an increase in the electric resistance and the generation of the Hall voltage transverse to the current flow in the conductor. The orbital motion corresponds to a redistribution of the electrons in the conduction band onto Landau cylinders in momentum space. At high magnetic fields this leads to a periodic oscillation of the electronic crystal properties as a function of the magnetic field. Experimental studies of these oscillations yield information about the extreme cross-sections of the Fermi surface perpendicular to the magnetic field direction. The de Haas–van Alphen effect is discussed. In the restricted geometry of a two-dimensional electron gas, new quantum effects appear, such as the integer and the fractional quantum Hall effect.

The brilliant American physicist Henry A. Rowland is perhaps best known for his mechanical fabrication of optical diffraction gratings, which were unique during his time and became famous as the "Rowland gratings". In the year 1870 he had completed his education as a civil engineer at the Rensselaer Polytechnic Institute (RPI) in Troy in the Federal State of New York. After some time as an Assistant Professor for science at the Wooster University in Ohio, he returned in 1872 to the RPI with an appointment as Instructor of Physics. During the winter semester 1875/1876 he visited the Institute directed by Hermann von Helmholtz in Berlin. Rowland was particularly interested in the theory of electricity and the field of "electrodynamics", founded by Michael Faraday and James Clerk Maxwell in England. During his journey to Berlin, Rowland passed through Cambridge in England where he visited Maxwell. As a guest of Helmholtz, in an astonishingly short time Rowland was able to demonstrate experimentally that electrically charged objects are accompanied by a magnetic field, if they move at a high velocity. In his report to the Berlin Academy during March 1876, Helmholtz himself declared: "*Mr. Rowland has just performed a series of direct experiments in the physics laboratory of this university, which present the positive proof, that the motion of electrically charged objects [aside from conductors!] is also electro-*

© Springer Nature Switzerland AG 2019
R. P. Huebener, *Conductors, Semiconductors,
Superconductors*, Undergraduate Lecture Notes in Physics,
https://doi.org/10.1007/978-3-030-31420-0_7

magnetically active." Such fundamental experiments in the field of electrodynamics were then at the frontiers of physics. Following the Semester in Berlin, Rowland went to the American Johns Hopkins University which had just been founded in Baltimore in the Federal State of Maryland.

Having returned to the USA and continuing his previous research activities, Rowland was interested to find out if an electric current in a metallic conductor is deflected sideways by an external magnetic field, thereby causing an additional electric voltage signal perpendicular to the direction of the electric current. After Rowland himself could not detect any effect, he turned this, for the time quite ambitious measurement task, over to his student Edwin Herbert Hall. In the year 1879 Hall observed the effect. Subsequently, this phenomenon is referred to as the Hall effect.

7.1 Hall Effect

The Hall effect represents one of the simplest phenomena caused by moving electric charge carriers, when an external magnetic field is also present. Electric charge carriers moving in a magnetic field experience a force, which is oriented perpendicular to both the direction of their motion and the direction of the magnetic field. This force, named after the Dutch theoretical physicist Hendrik Antoon Lorentz, is

$$\mathbf{f}_{\mathrm{L}} = q\,\mathbf{v} \times \mathbf{B} \tag{7.1}$$

It is proportional to the magnetic field $\mathbf{B}$, and also proportional to the moving electric charge q and to the component of the velocity $\mathbf{v}$ of the charge carriers perpendicular to the magnetic field. Reversing the sign of each of these three factors leads to a sign reversal of the Lorentz force. The Lorentz force vanishes, if the charge carriers move parallel to the direction of the magnetic field such that the velocity component perpendicular to the magnetic field remains zero. As a result of the Lorentz force, the motion of the free charge carriers along a straight line is deflected into a helical or circular trajectory (Fig. 7.1a). In the case of the electric current density (5.3) $\mathbf{j} = n(-e)\Delta\mathbf{v}_{x}$ along x-direction in a conductor, and in the presence of a magnetic field $\mathbf{B}$ along z-direction, the Lorentz force is oriented along y-direction: $\mathbf{f}_{\mathrm{Ly}} = q\,\Delta\mathbf{v}_{x} \times \mathbf{B}$. Here we have inserted the drift velocity $\Delta\mathbf{v}_{x}$ into (7.1). Because of the force $\mathbf{f}_{\mathrm{Ly}}$, the positive and negative electric charges accumulate at the opposite sides of the conductor, respectively, assuming that they move in the same direction (Fig. 7.1b), and thereby they generate an electric field $\mathbf{E}_{y}$ along y-direction. In the stationary state, the Lorentz force $\mathbf{f}_{\mathrm{Ly}}$ is compensated by the electrostatic force $q\,\mathbf{E}_{y}$, and one finds

Fig. 7.1 Lorentz force acting on electric charges moving in a magnetic field. The magnetic field is directed perpendicular to the plane of the paper. **a** Because of the Lorentz force, the motion of electric charges perpendicular to the direction of the magnetic field is deflected onto a circular orbit. **b** Hall effect: During electric current flow in a conductor, the sideways deflection of the current due to the Lorentz force causes the accumulation of charges with opposite sign, respectively, on both sides of the conductor (assuming that the opposite charges move in the same direction). This results in an electric voltage perpendicular to the main current direction and perpendicular to the direction of the magnetic field. **c** Magneto-resistance: The deflection of the electric charges into circular orbits because of the Lorentz force hinders the electric current flow along its main direction and causes an increase in the electrical resistance. The crosses mark the locations, where the electrons experience a collision process

$$q\,\Delta\mathbf{v}_x \times \mathbf{B} = q\,\mathbf{E}_y. \tag{7.2}$$

From this follows

$$\left|\mathbf{E}_y\right| = \left|\Delta\mathbf{v}_x \times \mathbf{B}\right| = \frac{1}{(-e)n}j\,B \equiv R_H\,j\,B. \tag{7.3}$$

$\mathbf{E}_y$ denotes the electrical Hall field, characterizing the Hall effect. In (7.3) we have used $q = -e$. The coefficient $R_H = \frac{1}{(-e)n}$ is the Hall constant (which is negative in the case $q = -e$). We see, that the Hall constant yields information about the concentration of the moving charge carriers, and that the sign of the electrical Hall field indicates the sign of these charge carriers. However, early on occasionally one had observed also a positive sign of the Hall constant, which remained unexplained (anomalous Hall effect) until the concept of the (positive) holes in the electronic band structure was established.

At this point it is important to note, that for the same direction of the electric current, positive and negative electric charges move in opposite directions. Hence, in the case of reversal of the sign of the moving charges, the sign of the Lorentz force changes twice, such that in the end it remains the same. Therefore, the simultaneously moving positive and negative electric charges are driven toward the same side of the current-carrying conductor. As we have discussed in Chap. 6 in the context of the hole concept, already early on the Hall effect indicated frequently, that the moving charge carriers act like (positive) holes, which originate from the region near the upper edge of an almost completely filled energy band. In addition to the sign of the moving charges, from the Hall effect one can also determine the concentration of the moving charge carriers in the electrically conducting material.

In principle, the behavior we have discussed so far is expected in the same way for both metals and for semiconductors, as long as there exists only a single kind of charge carrier. However, if the electric current is transported by two or more different kinds of charge carriers, the situation can become complicated very rapidly. For example, the Hall effect completely disappears, if simultaneously positive and negative charge carriers are present with exactly equal concentration and if they also contribute to the electric current with the same mobility. Since in the magnetic field the positive and negative charge carriers are driven to the same side of the current-carrying conductor, there the charges with the opposite sign compensate each other exactly in this case, such that no Hall effect remains.

An impressive example of the Lorentz force can be observed in the phenomenon of the northern lights. It is caused by the impact of electrically charged particles, particularly of protons, originating from the sun. In the earth's magnetic field the particles are deflected to higher latitudes along circular trajectories, where they optically excite the gas molecules at about 100 km altitude. Also in the large accelerators the electrically charged particles are kept on their proper trajectories by means of magnetic coils and the Lorentz force. Furthermore, this force and the effected deflection of the trajectory of electrically charged particles represents the principle of the "magnetic bottle", which is supposed to keep hot plasma sufficiently far away from the reactor walls of the long-term-project nuclear fusion reactors. Finally, it is also the same force which acts upon an electrical conductor during electric current flow and which represents, for example, the principle of the electric motor.

7.2 Magneto-Resistance

The deflection of the electric charge carriers onto helical or circular orbits in a magnetic field by means of the Lorentz force also affects the electric resistivity (Fig. 7.1c). This effect is referred to as the magneto-resistance. On general grounds we expect an increase in resistance due to the magnetic field, since the electric current flow is hindered, if the charges are forced to follow helical or circular orbits, instead of moving only in a single direction. At not too high magnetic fields, the resistance increases with magnetic field proportional to B^2 (This B^2 dependence is expected, since the increase of resistance must be independent of the sign of the magnetic field). The increase of the electrical resistance depends on whether, and in which way, both electrons and holes contribute to the electric current. In single crystals the resistance increase in a magnetic field can also depend strongly on the crystallographic direction. In recent years magneto-resistance effects, in which the angular momentum or spin of the electrons plays a central role, have received a large amount of attention. We will return to these spectacular developments in Chap. 10 when we discuss the subject of the "giant magneto-resistance" and spintronics.

7.3 Landau Theory, Landau Cylinders, de Haas–van Alphen Effect

The exact quantum mechanical theory of electrons in the conduction band of a metal in the presence of a strong magnetic field, is due to the Russian Lew Dawidowitsch Landau. In the year 1930 at the age of only 22 years he published his famous paper on the diamagnetism of metals. One year earlier, following his graduation in Leningrad, Landau had begun a two-year visiting tour to European Research Institutes, which brought him among other places to Zurich, Copenhagen, Cambridge, Berlin, and Leipzig. In his paper Landau showed that the energy spectrum of the electrons is strongly modified by the magnetic field. In Chaps. 4 and 5 we have discussed, how the energy spectrum of the electrons in the conduction band of a metal is described in terms of the wave vectors which characterize the electron matter waves during their propagation along all three spatial directions. However, because of the magnetic field and the Lorentz force, the electron motion perpendicular to the direction of the magnetic field is forced into a circular orbit. Assuming that the magnetic field is oriented along the direction of the z-coordinate, the circular orbit is located within the plane of the x- and y-coordinate. Whereas in the absence of a magnetic field (B = 0) the energy spectrum of the electrons is given by

$$\varepsilon = \frac{\hbar^2}{2m_e}\left(k_x^2 + k_y^2 + k_z^2\right); \quad B = 0 \tag{7.4}$$

in the presence of a magnetic field $B = B_z \neq 0$ oriented along z-direction it is

$$\varepsilon = \hbar\omega_c\left(\ell + \frac{1}{2}\right) + \frac{\hbar^2}{2m_c}k_z^2; \quad B = B_z \neq 0. \tag{7.5}$$

Here, ℓ is an integer, and m_c is the cyclotron mass of the electrons.

$$\omega_c = \frac{e\,B}{m_c} \tag{7.6}$$

is the cyclotron angular frequency of the electrons. We see, that during their circular orbit in the x-y plane, the energy of the electrons is quantized in the unit of the cyclotron energy $\hbar\omega_c$. The wave vectors within the x-y plane are no longer relevant, since the corresponding states form new combined states with orbital motion. Only the electron motion along the z-direction remains unchanged in the form of the corresponding matter wave defined by the wave vector k_z along the z-direction. In this context we remember that the Lorentz force vanishes if the electrons move parallel to the direction of the magnetic field.

In the absence of a magnetic field, the three-dimensional **k**-space of the wave vectors is evenly filled with states to be occupied. On the other hand, in the case of an existing magnetic field, because of the energy quantization (7.5) according to Landau, the allowed states are redistributed into a series of coaxial "Landau cylinders". Now the states which can be occupied are restricted to these cylinders. The axis of the cylinders is oriented along the direction of the k_Z wave vector, i.e., along the same direction as that of the magnetic field. The energetic distance between two subsequent Landau cylinders is $\hbar\omega_c$, which increases proportional to B. Hence, at high magnetic fields, their energetic distance is relatively large. At a magnetic field B = 1 T one obtains a typical value $\hbar\omega_c \approx 10^{-4}$ eV. Figure 7.2 shows an example.

For the manifestation of the energy quantization according to Landau it is necessary, that the circular orbits of the electrons in the magnetic field are not disrupted by collision processes experienced by the electrons. The circular orbits should be completely traversed without perturbation at least one time. Since the number of the collisions, for example with phonons, strongly decreases with decreasing temperature, in addition to high magnetic fields, temperatures as low as possible are required for the experimental observation of the quantum structure associated with the Landau cylinders. Furthermore, sufficiently low temperatures also ensure, that the thermal energy $k_B T$ is distinctly smaller than the energetic distance $\hbar\omega_c$ between two neighboring Landau cylinders, and that the quantum structure is not smeared out because of the thermal energy $k_B T$. Finally, single-crystals with the highest possible purity should be used for the experiments.

Because of these reasons, for the distinct manifestation of the Landau cylinders, the following conditions must be satisfied (where τ denotes the scattering time):

$$\omega_c \tau > 1 \tag{7.7}$$

and

$$k_B T < \hbar\omega_c. \tag{7.8}$$

Fig. 7.2 In high magnetic fields the energy quantization according to Landau leads to a redistribution of the states, which can be occupied by the electrons in **k**-space, onto the walls of a family of coaxial cylinders. The common axis of the cylinders is oriented along the direction of the magnetic field. The distance between the cylinder walls in **k**-space increases proportionally with the magnetic field

Already by 1930 Landau had recognized that, because of the energy quantization of the conduction electrons in a magnetic field discussed by himself for the first time, macroscopic material properties, such as for example diamagnetism, should display an exactly periodic oscillation as a function of the magnetic field. However, he felt that the necessary purity criteria could not be reached for the available sample materials, and that, hence, the effect would remain unobservable. Since the energy difference $\hbar\omega_c$ between two neighboring Landau cylinders increases proportional to B, the number of the Landau cylinders to be occupied up to the Fermi energy with increasing magnetic field becomes smaller and smaller. In the case of a monotonically increasing magnetic field, the redistribution of the electrons onto the Landau cylinders according to the energy spectrum (7.5) leads to a periodic oscillation of the total energy of the electrons. At $\varepsilon_F = \hbar\omega_c \ell$ the total energy of the electrons passes through a minimum, and at $\varepsilon_F = \hbar\omega_c(\ell + \frac{1}{2})$ through a maximum. As a consequence, this also results in oscillations of the other electronic sample properties as a function of the magnetic field, such as, for example, diamagnetism.

Contrary to the apprehension by Landau, the oscillations of diamagnetism were observed for the first time by the year 1930 in bismuth single crystals by the Dutch Wander Johannes de Haas and Pieter M. van Alphen in Leiden. The effect is now referred to as the de Haas–van Alphen effect. At the time Rudolf E. Peierls contributed significantly to its further theoretical clarification. An important theoretical advance originated from Lars Onsager during a visit to Cambridge, England, in the Academic Year 1950/1951. Strongly emphasizing the geometrical interpretation of the Fermi surface in the three-dimensional **k**-space of wave vectors, he showed that only the extreme cross-sections of the Fermi surface taken perpendicular to the direction of **B** contribute to this effect. The contributions of all other parts of the Fermi surface are irrelevant, since they cancel each other. The period of the de Haas–van Alphen oscillations is inversely proportional to the extreme cross-section of the Fermi surface, i.e., inversely proportional to the largest and to the smallest cross-section (Here the extreme cross-sections are taken perpendicular to the direction of the magnetic field).

For a more detailed discussion we return to the cyclotron mass m_c and the cyclotron frequency ω_c introduced in (7.5) and (7.6), respectively. Writing the inverse cyclotron frequency in the form

$$2\pi/\omega_c = \oint \frac{d\mathbf{k}}{|\dot{\mathbf{k}}|} \tag{7.9}$$

and using

$$|\dot{\mathbf{k}}| = \frac{e}{\hbar}\mathbf{v} \times \mathbf{B} \tag{7.10}$$

based on the Lorentz force (7.1), writing q = e, and introducing the velocity component $v_\perp$ perpendicular to the magnetic field, we obtain

$$2\pi/\omega_c = \frac{\hbar}{e\,B} \oint \frac{d\mathbf{k}}{\mathbf{v}_\perp(\mathbf{k})} = \frac{2\pi\,m_c}{e\,B} \tag{7.11}$$

and finally

$$m_c = \frac{\hbar}{2\pi} \oint \frac{d\mathbf{k}}{\mathbf{v}_\perp(\mathbf{k})} \tag{7.12}$$

Here $d\mathbf{k}$ is a line element along the electron orbit in $\mathbf{k}$-space. With $v_\perp = (1/\hbar)(d\varepsilon/dk_\perp)$ we have $m_c = \frac{\hbar^2}{2\pi} \oint d\mathbf{k} \frac{\Delta k_\perp}{\Delta\varepsilon}$. Writing $\oint d\mathbf{k} \frac{\Delta k_\perp}{\Delta\varepsilon} = \frac{1}{\Delta\varepsilon} \oint d\mathbf{k}\, \Delta k_\perp$ we note that the integral $\oint$ on the rhs represents the difference ΔA of the areas A enclosed by the electron orbits at the energies ε and $\varepsilon + \Delta\varepsilon$ in $\mathbf{k}$-space, respectively. Finally, we obtain

$$m_c = \frac{\hbar^2}{2\pi}\left(\frac{dA}{d\varepsilon}\right), \quad k_\| = \text{const} \tag{7.13}$$

We see that the cyclotron mass m_c contains the energy dependence of the area A enclosed by the electron orbit in $\mathbf{k}$-space.

The quantization of the orbits in $\mathbf{k}$-space and of the magnetic flux are equivalent: with $\Delta\varepsilon = \hbar\,\omega_c = \hbar\,e\,B/m_c$ and $\Delta A = (dA/d\varepsilon)\Delta\varepsilon$, inserting (7.13), we find

$$A_\ell = (2\pi eB/\hbar)(\ell + \frac{1}{2}) \tag{7.14}$$

The area A_ℓ in $\mathbf{k}$-space is related to the area F_ℓ in regular space by

$$F_\ell = (\hbar/eB)^2 A_\ell \tag{7.15}$$

(This can be seen from (7.10) by writing $\dot{\mathbf{k}} = \Delta\mathbf{k}/\Delta t$ and $v_\perp = \Delta r/\Delta t$). For the magnetic flux Φ_ℓ passing through the area F_ℓ we find from (7.14) and (7.15)

$$\Phi_\ell = B\,F_\ell = (h/e)(\ell + \frac{1}{2}) \tag{7.16}$$

The quantity $(h/e) = 1.034 \times 10^{-15}$ Vs represents the magnetic flux quantum in a non-superconductor. We will deal with it again in the following Section.

As a function of the magnetic field $\mathbf{B}$ the de Haas–van Alphen oscillations show an exact periodicity when plotted versus $\mathbf{B}^{-1}$. From (7.14) we see that each Landau level contributes the amount $\Delta A = (2\pi eB)/\hbar$ to the total cross section of the Fermi surface, and we have $A = (2\pi eB/\hbar)\lambda$, where λ is an integer. Writing $1/B_o = (2\pi e/\hbar A)$, we obtain the periodicity $B_o/B = \lambda$.

In the three-dimensional case only the extremal cross-sections of the Fermi surface perpendicular to $\mathbf{B}$, i.e., the maximal A_{max} and the minimal cross-sections

A_{min}, are relevant. As examples we mention the belly and the neck orbits on the Fermi surface of copper shown in Fig. 5.3. So we find the periodicity in $1/B$

$$\frac{1}{B_o} = \frac{2\pi e}{\hbar} \frac{1}{A_{max}} \quad \text{or} \quad \frac{1}{B_o} = \frac{2\pi e}{\hbar} \frac{1}{A_{min}} \tag{7.17}$$

The de Haas–van Alphen effect is connected with the magnetization M. With the free energy $F = U - TS$ the magnetization is given by $M = -\frac{1}{V}\left(\frac{\partial F}{\partial B}\right)_{T,V}$. In the case $T \approx 0$ we have

$$M = -\frac{1}{V}\left(\frac{\partial U}{\partial B}\right)_{T,V} \tag{7.18}$$

(U = inner energy, S = entropy, V = volume). The periodic oscillation of the energy of the electrons mentioned above leads to a corresponding oscillation of M with the period $1/B_o$ when plotted versus $1/B$.

Subsequently, it transpired that Ilya Mikhailovich Lifshitz in Moscow had developed similar ideas independent of Onsager. In Cambridge, David Shoenberg, we had mentioned in Chap. 5, and his collaborators have utilized these theoretical ideas particularly successfully.

The de Haas–van Alphen effect then has become an important experimental tool, particularly in the 1950s and 1960s, for determining the shape of the Fermi surface in many materials, as long as the materials could be prepared in the form of sufficiently pure single crystals. Eventually, impressively fine details were discovered, and the experimental techniques were continuously improved. Large deviations of the Fermi surface from a simple spherical shape were observed. In some materials, as for example the multi-valency metals, the Fermi surface often consists of half a dozen or more separate parts, which are associated either with electrons or with holes. For such parts, notations such as monster, cap, lense, butterfly, needle, or cigar were invented. The belly and the necks of the Fermi surface of copper, discovered by Alfred Brian Pippard and shown in Fig. 5.3, represent only the first and, still relatively simple, example.

As one would expect, similar oscillations as in the de Haas–van Alphen effect also appear in other physical material properties, which are influenced by the mobile electric charge carriers. The "Shubnikov–de Haas oscillations" of the electric conductivity represent one example. Even in the chemical reaction rate on the surface of metallic catalysts, during the variation of the magnetic field, oscillations have been observed which have the same origin.

For our further discussion it is useful to look more closely at the density of states which can be occupied by electrons as a function of the electron energy. We start with the three-dimensional case. In this case, which is usually applicable, according to (6.3) and (6.7), the density of the possible energy levels increases proportionally with the square-root of the electron energy as long as no magnetic field is present. However, if a magnetic field exists, the redistribution of the energy levels onto the Landau cylinders in **k**-space causes the curve to be superimposed by many sharp peaks following each other within the energetic distance $\hbar\omega_c$ of the cyclotron energy. On the other hand, in the two-dimensional case, in the absence of a magnetic field, we find that the density of the energy levels, which can be occupied, is independent of the energy, i.e., it is constant.

In Fig. 7.3 we present an overview of the energy dependence of the density of states at zero magnetic field in the four cases of a different number of spatial dimensions, from three (top) to zero (bottom). In the one-dimensional case, the

Fig. 7.3 Dependence of the density of states upon the energy E at zero magnetic field in the four cases of a different number of spatial dimensions from three (top) to zero (bottom). B = 0

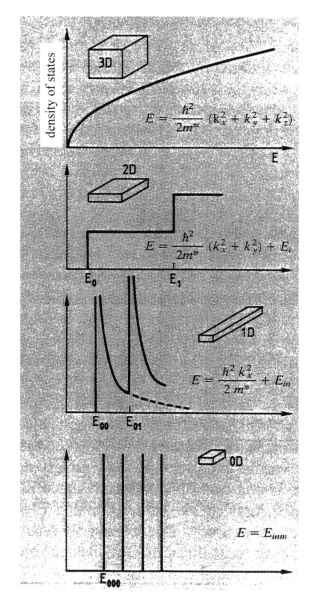

$$E = \frac{\hbar^2}{2m^*} (k_x^2 + k_y^2 + k_z^2)$$

$$E = \frac{\hbar^2}{2m^*} (k_x^2 + k_y^2) + E_i$$

$$E = \frac{\hbar^2 k_x^2}{2\,m^*} + E_{in}$$

$$E = E_{inm}$$

density of states decreases from one energy level E_{in} to the next higher one with $(1/energy)^{1/2}$, whereas in the zero-dimensional case there only exist the discrete energy levels E_{inm}. In the two-dimensional case, the independence of the density of states of the energy, at high magnetic fields, leads to novel physical properties such as, for example, the quantum Hall effect. Next we wish to address these questions.

7.4   Integer Quantum Hall Effect

Based on the periodic boundary conditions, in Chap. 6 we have shown, that in the one-dimensional case the density of states $w(\mathbf{k})$ in $\mathbf{k}$-space is $w_1(\mathbf{k}) = L/2\pi$, where L denotes the physical dimension of the crystal. By extension to two dimensions, one obtains $w_2(\mathbf{k}) = (L/2\pi)^2$. In the two-dimensional case, in analogy to (6.6), the density of states *per energy interval* (using (4.13) and ignoring the spin of the electrons) is given by

$$D_2(\varepsilon) = w_2(\mathbf{k}) \frac{\partial(\mathbf{k} - \text{space area})}{\partial \varepsilon} = w_2(\mathbf{k}) 2\pi k \frac{\partial k}{\partial \varepsilon} = \left(\frac{L}{2\pi}\right)^2 2\pi k \frac{m}{\hbar^2 k} \qquad (7.19)$$

and $D_2(\varepsilon)$ per unit area is

$$D_2(\varepsilon)/L^2 = m/2\pi\hbar^2. \qquad (7.20)$$

In (7.20) the density of states is independent of the energy. Using the energy interval $\hbar\omega_c$ between two subsequent Landau levels (from (7.5) with $k_z = 0$), we find the number N of the states per area and per Landau level:

$$N = [D_2(e)/L^2]\hbar\,\omega_c = \frac{e\,B}{h}. \qquad (7.21)$$

As discussed above, in order to clearly observe the quantum structure according to Landau the conditions (7.7) and (7.8) must be satisfied. This case is shown schematically in Fig. 7.4.

We return to the Hall effect (7.3), and in the two-dimensional case we have:

$$E_y = \frac{1}{(-e)n_2} j_2\,B. \qquad (7.22)$$

Here $j_2 = I/w$ denotes the current density in a two-dimensional conductor of width w carrying the current I. The quantity n_2 is the number of electrons per area. If the Fermi energy ε_F is located exactly between two Landau levels, all Landau levels below (above) ε_F are occupied (unoccupied). In this case n_2 in (7.22) amounts to:

Fig. 7.4 Density of states per area $D_2(\varepsilon)/ L^2$, plotted versus the normalized energy $\varepsilon/ \hbar \omega_c$

$$n_2 = z\,N = z\frac{e\,B}{h}, \tag{7.23}$$

where z is an integer number (Frequently, z is referred to as filling factor). For the Hall resistance R_{xy} one obtains:

$$R_{xy} = \frac{E_y}{j_2} = \frac{1}{z}\frac{h}{e^2}. \tag{7.24}$$

The Hall resistance R_{xy} only depends upon the fundamental constants h and e, an important and surprising result.

In order to discuss the quantum Hall effect, we consider a two-dimensional crystal (two-dimensional electron gas) in a high magnetic field, the field being oriented perpendicular to the plane in which the crystal is located. Now the Landau cylinders are reduced to Landau circles obtained as a two-dimensional cut through the coaxial cylinders perpendicular to the cylinder axis. The sequence of the Landau circles, placed within each other around their common center, again corresponds exactly to the quantized energy of the electrons in their circular orbits within the plane of the crystal, representing multiples of the cyclotron energy $\hbar\omega_c$. The constant density of the energy levels of the two-dimensional crystal, which can be occupied by electrons, according to (7.21) leads to the consequence that all energy intervals (of magnitude $\hbar\omega_c$) between the subsequent Landau circles contain exactly the same number of energy levels to be occupied. Therefore, the energy levels corresponding to the individual Landau circles are also occupied by exactly the same number of electrons. The energy spectrum of the electrons displays a sequence of sharp and exactly equal peaks, appearing periodically along the energy axis with a distance given by the cyclotron energy $\hbar\omega_c$ (Fig. 7.4). If it is possible to increase continuously the number of the mobile electrons in the two-dimensional crystal, then the electrical properties should change in a step-wise manner each time an additional Landau level just becomes filled up with electrons.

More than 30 years ago the German Klaus von Klitzing was interested in such effects in the context of his research at the Physical Institute of the University of Würzburg. In the Fall of 1979 he went to the German-French High-Magnetic-Field Laboratory in Grenoble for a research visit, since in Grenoble magnetic coils were available for stronger fields than in Würzburg. At the time, the previous thesis advisor of Klaus von Klitzing, Gottfried Landwehr, was in charge of the Magnet

Laboratory in Grenoble. During his experiments in Grenoble von Klitzing used a field-effect transistor made from silicon, provided to him by the Siemens Company. This device represents one of the many further developments following the early transistor types. Near the semiconductor surface the mobile charge carriers are confined to a narrow two-dimensional region. The semiconductor surface is covered by a thin, electrically insulating layer of silicon oxide (SiO_2), on the other side of which a metal electrode is attached. Between the metal electrode and the silicon crystal an electric voltage, referred to as the gate voltage, can be applied. This gate voltage allows one to vary continuously the concentration of the mobile charge carriers within their two-dimensional confined region near the silicon surface. With this arrangement the experimental requirements for the observation of the step structure discussed above appear to be well satisfied.

On the night of 4th of February 1980, von Klitzing discovered that, in a high magnetic field and at the low temperature of 1.5 K, the Hall resistivity (measured perpendicular to the electric current) of his field-effect transistor displayed particularly sharp and regular steps as a function of the gate voltage. On the other hand, the electrical resistivity measured along the direction of the current showed strong oscillations as a function of the gate voltage and dropped down to zero at each horizontal step of the Hall resistivity. All steps and the oscillations disappeared, if the magnetic field was turned off. On the same night, von Klitzing had already recognized, that the steps represent something fundamental which depends only upon two fundamental physical constants, and which is exactly quantized (Fig. 7.5). With increasing gate voltage the Landau levels are filled sequentially with mobile charge carriers. Simultaneously, the Hall resistance decreases. However, this decrease is always interrupted and an exactly constant intermediate resistance value appears, if a Landau level has just been filled up, and if the following level cannot yet be reached. In this way the exactly quantized values of the Hall resistance $(1/z)$ (h/e^2) appear, which von Klitzing observed on his measured curve (Fig. 7.6). Here z is an integer number, such as 2, 3, 4, etc. The quantity h is Planck's constant, and e is the charge of an electron. For the unit of the quantized Hall resistance, von Klitzing obtained the value $h/e^2 = 25813\,\Omega$. He had succeeded in the pioneering discovery of the quantum Hall effect.

From (7.23) for the number n_2 of the electrons per area and per Landau level we find an important connection with the number of magnetic flux quanta (h/e), if we multiply the number n_2 with the sample area F. One obtains

$$n_2 F = z \frac{e\,B}{h} F = z \frac{\Phi}{h/e} \qquad (7.25)$$

where Φ denotes the magnetic flux (B F) passing through the sample. This means that, at the exactly quantized resistance steps, each magnetic flux quantum in the sample is connected with the same number z of electrons. Again, we will find a similar important relation, between the number of electrons and the number of magnetic flux quanta per unit area, in the case of the fractional quantum Hall effect, which we will discuss in the following section.

Fig. 7.5 Entry of Klaus von Klitzing in his work notebook on February 4, 1980, the day on which he discovered the quantum Hall effect (K. von Klitzing)

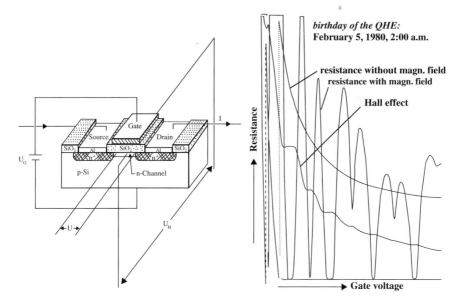

Fig. 7.6 Integer Quantum Hall effect: Electrical resistance and Hall resistance in a high magnetic field and at a temperature of 1.5 K, plotted as a function of the gate voltage for the two-dimensional electron gas of the field-effect transistor made from silicon shown on the left, in which early on February 5, 1980 Klaus von Klitzing discovered the quantum Hall effect. The smooth resistance curve without any steps was observed in the absence of a magnetic field (Klaus von Klitzing)

From the very first moment it was clear that the quantized value of the electrical resistance in the quantum Hall effect provided an excellent opportunity for a new quantum definition of the unit of the electrical resistance. Soon the German Bureau of Standards (*Physikalisch-Technische Bundesanstalt*) in Braunschweig, as well as the National Standards Bureaus in other countries had taken this opportunity. Since January 1, 1990 the "von-Klitzing-constant" h/e^2 has represented the legal definition of the unit of the electrical resistance based on the quantum Hall effect. Also, the accuracy of the determination of the von-Klitzing-constant was improved further, and the official value today is $h/e^2 = 25812.807 \ \Omega$. In Chap. 11 we will return to the quantum Hall effect in connection with the discovery of the properties of graphene.

However, von Klitzing was not the first to have observed step-like structures in the Hall resistance and oscillations of the electrical resistance along the current direction as a function of the gate voltage in a field-effect transistor. In Tokyo a few years earlier the Japanese group of S. Kawaji had obtained similar, but not so clearly expressed curves as von Klitzing. Furthermore, this group did not notice the fundamental importance of their results in terms of a quantized electrical resistance value depending only on two fundamental physical constants.

Because of the quantum Hall effect, the two-dimensional electron gas on the surface of a semiconductor has become very famous. Even more than 30 years after the discovery of the effect the theoretical discussion is by no means closed. Based on the idea of the filling of the Landau levels with increasing gate voltage, the quantized values of the Hall resistance $(1/z)$ (h/e^2) can be quickly derived, but all the details of the measured curves are still not yet completely theoretically explained.

7.5 Fractional Quantum Hall Effect

Because of the permanent progress in the preparation of semiconductor materials, the physics of the two-dimensional electron gas in high magnetic fields also received a strongly increasing amount of attention. Here the search for the "Wigner crystal" generated a lot of activity. In 1938 Eugene Paul Wigner had already predicted theoretically, that at sufficiently low temperatures electrons would arrange themselves in a perfectly ordered crystal lattice, if they are confined to a two-dimensional space, as for example on the surface of a semiconductor. The field-effect transistor based on silicon still appeared insufficient in its quality for an experimental study of this phenomenon of crystallization. However, the situation improved considerably, when near the end of the 1970s modulation-doped single-crystalline layers of semiconductors could be fabricated. Modulation doping of semiconductors is based on the obvious idea to spatially separate the mobile electrons from the donor atoms from which they originate. In this way one can achieve, in particular at low temperatures, that the electrons propagate through the semiconductor at high speed and without collisions with the ionized donors. Therefore, this type of "hetero-structure" promised to yield particularly fast and low-noise transistors, such that many laboratories worldwide then concentrated on this development. In this context, mainly single-crystalline layers of the III–V semiconductor galliumarsenide (GaAs), in combination with a galliumarsenide layer modified by an admixture of aluminum ($Al_xGa_{1-x}As$), were interesting. For example, silicon donor atoms (with four electrons in the outer shell) are implanted precisely at the locations of the gallium and aluminum atoms (with only three electrons in the outer shell, respectively) within the $Al_xGa_{1-x}As$ layer. Then the excess electrons of the silicon donors are transferred into the energetically lower conduction band of the adjoining GaAs layer, where they can propagate relatively freely.

Based on these latter materials, in 1978/1979 for the first time the preparation of a two-dimensional electron gas at the interface between GaAs and $Al_xGa_{1-x}As$ was accomplished. The pioneering work for the preparation of single-crystalline semi-conductor layers with nearly atomic accuracy was performed by the German Horst Ludwig Störmer together with his American colleagues Arthur C. Gossard and Raymond Dingle at the Bell Laboratories, and also by Gerhard Abstreiter and Klaus Ploog at the Max Planck Institute of Solid State Research in Stuttgart. Daniel Tsui, born in China and also working at the Bell Laboratories at the time, soon persuaded his colleague Horst Störmer to carry out electrical measurements on the new and highly promising semiconductor layers at the highest possible magnetic fields and at the lowest possible temperatures. Both felt that the Francis Bitter High-Magnetic-Field Laboratory at the famous MIT in the Federal State of Massachusetts would be particularly suitable for such experiments. At this laboratory, magnetic fields up to about one million times higher than the earth's magnetic field could be generated with electric coils. Here Tsui and Störmer performed their experiments, in which they varied the magnetic field while the density of the two-dimensional electron gas at the interface of their hetero-structure sample was kept constant. After cooling down to about 2 K, as expected, they observed the horizontal steps of the Hall resistance, already well known from the quantum Hall effect. However, after they had cooled the sample further down to below 0.5 K, in the highest range of the magnetic field they discovered something completely new: now a step appeared at the Hall resistance $3(h/e^2)$, i.e., at z = 1/3, if we express the Hall resistance in the form $(1/z)$ (h/e^2) which we have used above. During the following years additional plateaus of the Hall resistance were found, with other fractional values of z, such as 1/3, 2/3, 2/5, 3/5, 3/7, 4/7, etc. (Fig. 7.7). In each case the Hall resistances $(1/z)$ (h/e^2) with the fractional values of z appeared with exactly the same precision as in the quantum Hall effect with the integer values of z. Similar to the latter effect, Tsui and Störmer observed that the electrical resistance (measured along the direction of the current) also dropped down to values near zero each time a horizontal plateau of the Hall resistance was reached. The discovery of Tsui and Störmer was subsequently referred to as the fractional quantum Hall effect. In contrast to this, the effect discovered by Klaus von Klitzing is called integral or integer quantum Hall effect.

As we have previously discussed, because of the Lorentz force the electrons are moving along circular orbits, if their motion occurs perpendicular to the magnetic field. During the experiments mentioned above, the magnetic field was always oriented perpendicular to the plane of the two-dimensional electron gas, such that the circular orbits were also located within this plane. The diameter of the circular orbits is inversely proportional to the magnitude of the magnetic field. Hence, with increasing magnetic field the circular orbits contract, and eventually they reach a diameter which is smaller than the average distance between two neighboring electrons. In this case, at low temperatures, all electrons occupy only the lowest Landau level, and we are dealing with the "extreme quantum limit". (In this case we must remember, that the Landau level exists within **k**-space. In position space, all individual electrons of the conduction band are spatially separated from each other, and are degenerated with respect to their energy). On the other hand, the quantum mechanical wave function of the electrons must be single-valued at each coordinate point in the semiconductor. This requires, that the magnetic flux penetrating the two-dimensional electron gas is quantized in units of the magnetic flux quantum (h/e). The observation of the fractional quantum Hall effect indicates that the electrons prefer distinct distances from each other in their two-dimensional arrangement. At these distinct distances the ratio z of the number of electrons per

Fig. 7.7 Fractional quantum Hall effect in the two-dimensional electron gas of a modulation doped GaAs/AlGaAs semiconductor hetero-structure at a temperature of about 0.1 K. The electrical resistance R and the Hall Resistance R_H are plotted as a function of the magnetic field. The Hall resistance displays many plateaus for the indicated fractional values of z, if the quantized Hall resistance is written in the form $(1/z)$ (h/e^2) (Horst L. Störmer)

unit area and the number of magnetic flux quanta per unit area take up exactly only rational values such as the fraction of two integer numbers as indicated above. Magnetic flux quanta and electrons are then intimately connected with each other. Furthermore, the experimental observations suggest the existence of a gap in the energy spectrum of the electron system, similar as in the integral quantum Hall effect. However, in the present case the interaction between the electrons also appears to play an important role. At high magnetic fields the electrons, together with the magnetic flux quanta, seem to condense into a novel quantum liquid.

The American Robert Betts Laughlin, presently working at Stanford University in California, has proposed an amazingly simple manybody wave function for describing this new manybody groundstate, which can explain many aspects of the experimental results. In particular, Laughlin could account for the exclusively odd values of the numbers in the denominator of the rational values z for the ratio we have discussed above in terms of the required anti-symmetry of the total wave function. In the meantime, the experimental and theoretical treatment of the fractional quantum Hall effect has lead to new concepts about novel particles composed of magnetic flux quanta and electrons, which can appear as collective energetic excitations of the two-dimensional electron gas.

7.6 Generation of High Magnetic Fields

Before we conclude our discussion of the effects in high magnetic fields we wish to look a bit closer at the magnetic coils which were used, and at the corresponding technical developments. The fabrication of magnetic coils wound out of super-conducting wires began in the early 1960s. At that time an important progressive step took place in the production of the technically relevant superconductors. Since then one can find superconducting magnets in many laboratories, and the experiments at high magnetic fields have become much simpler than before. In order to reach the temperature range in which superconductivity occurs, the magnetic coils are cooled down to 4 K using liquid helium. The superconducting materials will be discussed in more detail in Chap. 8. Today, superconducting magnets generating magnetic fields up to about one million times higher than the earth's magnetic field are standard equipment for the relevant laboratories. By means of special con-struction measures, in the "hybrid magnets" the magnetic field can be increased further up to about twice this value. During recent years different centers have been established, mostly on a national scale, in which experiments at very high magnetic fields can be carried out. We have mentioned before the German-French High-Magnetic-Field Laboratory in Grenoble and the American Francis Bitter Laboratory at the MIT in Cambridge. As a continuation of the latter Laboratory, since a few years in the USA the National High-Magnetic-Field Laboratory at the Florida State University in Tallahassee in the Federal State of Florida has been operating. Further special facilities for high magnetic fields exist in Nijmegen in Holland as well as in Sendai and Tsukuba in Japan. In Hefei, China, a laboratory for static high magnetic fields is being built, which is operating from around 2013.

Already in the 1920s and 1930s there were laboratories in which experiments were performed in high magnetic fields. The Frenchman Aimé Cotton had con-structed a giant electromagnet near Paris, and the American Francis Bitter had built large electromagnets at the MIT in Cambridge. The Russian Pjotr Leonidovich Kapitza developed pulsed electromagnets in Cambridge, England. In all cases, in addition to the electric current, the consumption of cooling water was also enor-mously high, since the magnet coils were not yet fabricated from superconductors and generated a large amount of heat during their operation.

The Russian Pjotr Leonidovich Kapitza had studied at the Polytechnique Institute in St. Petersburg, where he was tutored by Abram Fedorovich Ioffe. In the year 1921 at the age of 27, he went to Ernest Rutherford in Cambridge, England, in order to learn more about current developments in physics. His career in Cambridge was highly successful, and in 1930 he became director of the newly established Mond Laboratory. At that time he was interested in strong magnetic fields, in order to deflect the tracks of alpha particles.

Therefore, he built a special pulse generator, with which he could generate, in a pulsed coil, the largest magnetic fields at that time. He was the first to employ this pulse technique to obtain high magnetic fields. In the meantime this method has been developed to a high level by different groups. Kapitza discovered, among other things, the linear increase of electrical resistance at high magnetic fields, a law which is named after him. At an early stage he turned to the subject of low temperature physics. When, during the Stalin era in 1934, he returned once again to his native Russia to visit his mother in Moscow, the authorities prohibited his return to England. Instead, they built a new Institute for him in Moscow. Later, his Institute for Physics Problems became highly famous. In 1937 Kapitza was able to attract the theoretical physicist Lew Dawidowitsch Landau from Kharkov in the Ukraine to his Institute in Moscow. During this time of political prosecutions in the then Soviet Union Landau was also arrested in the following year. Because of his personal intervention with Stalin, only after a whole year was Kapitza able to release Landau again from prison. Subsequently, Landau became the dominant father figure of Russian theoretical physics. The Russian experimental physicist Leo Vasilyevich Shubnikov, also from Kharkov and a close friend of Landau, was not so lucky as Landau, following his arrest. After a three-month detention, while awaiting trial, he received the death penalty and was shot on November 10, 1937. In 1931 Shubnikov had initiated the establishment of the first Low-Temperature Laboratory in the Soviet Union at the Ukrainian Physico-Technical Institute in Kharkov. Then, also at this location, experiments could be performed in the temperature range of liquid helium, similar to those in Leiden, Toronto, and Berlin. Shubnikov has distinguished himself primarily because of his pioneering research in the field of superconductivity. At an early stage, in his experiments he noted hints of a second kind of superconductors. This subject will be treated in Chap. 8.

Today, pulsed high magnetic fields, such as were used for the first time by Kapitza, are of extreme interest. Corresponding research facilities are operated in Los Alamos, USA, in Toulouse, France, and in Tokyo, Japan. In the German Research Center in Rossendorf near Dresden, since 2007 a new high-magnetic-field facility is in operation, in which pulsed magnetic fields up to 90 T can be generated (One Tesla is about 2×10^4 times the earth's magnetic field in central Europe). For the required electric current source the worldwide largest condenser bank was specially developed. In the case of the highest magnetic field, the pulse duration amounts to 11 ms. In the case of somewhat lower magnetic fields the pulse duration is longer. In Wuhan, China, a similarly powerful laboratory for pulsed magnetic fields is expected to be completed by 2013.

Chapter 8
The Winner: Superconductors

Abstract After Heike Kamerlingh Onnes succeeded in extending the range of experiments to much lower temperatures by the liquefaction of the noble gas helium, in 1911 he discovered superconductivity, where electric current flows without detectable resistance. Superconductivity requires that distinct critical values of the temperature and magnetic field are not exceeded. Eventually it was found that a magnetic field is expelled from the interior of a superconductor, referred to as Meissner effect and representing a fundamental property of superconductors. In the mixed state, a type-II superconductor is intersected by the Abrikosov lattice of magnetic flux quanta. Magnetic flux quantization and the Josephson effect are discussed. The first microscopic theory, the BCS theory, explains superconductivity in terms of a macroscopic quantum state formed by pairs of electrons (Cooper pairs) attracted to each other because of their interaction with phonons. The motion of the magnetic flux quanta, caused by the Lorentz force, represents the mechanism limiting the current flowing without resistance.

Heike Kamerlingh Onnes (Fig. 8.1) had succeeded in liquefying the noble gas helium in Leiden and in this way was able to reach the then low-temperature record of 4 K (−269 °C). During cooling down to low temperatures, in the year 1911, he made a surprising discovery: below a distinct temperature the electrical resistance of metals can vanish completely and cannot be detected experimentally. For the first time the phenomenon of "superconductivity", as it was afterwards called, had been observed.

After Kamerlingh Onnes had extended his experiments to the newly accessible range of distinctly lower temperatures than were possible before, he was also interested among other things in the question of how the electrical resistance of metals changes at these low temperatures. At the time there existed three different predictions about the behavior of the electrical resistance at low temperatures with decreasing temperature: (1) The resistance decreases down to the value zero, (2) it remains constant, (3) it increases again. Mercury as a metal appeared to be highly favorable for such measurements, since it can be prepared reasonably well with high purity, because of its low melting point (at room temperature it is already a liquid). The pioneering study had to be carried out with a material containing as few perturbing impurities as

© Springer Nature Switzerland AG 2019

R. P. Huebener, *Conductors, Semiconductors,*
Superconductors, Undergraduate Lecture Notes in Physics,
https://doi.org/10.1007/978-3-030-31420-0_8

Fig. 8.1 The Dutch physicist Heike Kamerlingh Onnes. In 1908 in Leiden for the first time he liquefied the noble gas helium. Three years later he discovered superconductivity (*Photo Kamerlingh Onnes Laboratory, University of Leiden*)

possible. Therefore, for one of the initial measurements a thin glass capillary filled with mercury was used. On April 8, 1911 Heike Kamerlingh Onnes, together with his assistants Cornelis Dorsman, Gerrit Jan Flim, and the student Gilles Holst, during cooling of the capillary filled with mercury, observed, how the electrical resistance of the sample decreased with decreasing temperature. However, when the temperature finally reached 4 K, the curve showed a sharp break, and the resistance dropped down abruptly to a small value which remained undetectable (Fig. 8.2). At first, there were some irritations, since it was presumed that the electric circuit of the measuring arrangement was defective, and that a short-circuit possibly caused the abrupt drop of the electrical resistance. However, after everything had been carefully checked, eventually it became clear, that the measuring technique was in order, and that a new phenomenon had been discovered. Later, the student Gilles Holst was employed by the N. V. Philips' Gloeilampenfabrieken in Eindhoven, and eventually he became the first director of Philips Research Laboratories.

Fig. 8.2 Discovery of superconductivity. Electrical resistance in ohms of a mercury sample plotted as a function of the temperature in Kelvin (H. Kamerlingh Onnes)

Following this first observation in mercury, superconductivity was also found in other metals such as, for example, in aluminum, lead, indium, tin, and zinc, as well as eventually also in alloys and metallic compounds. A compilation from the year 1969 lists about 350 different superconducting material systems. Superconductivity always appears only after cooling down below a characteristic temperature, the "critical temperature" T_C, having a specific value for each material. After the discovery of the high-temperature superconductors, which will be discussed in the following Chap. 9, the superconducting materials known up to then are referred to as classical superconductors. Of these classical superconductors the metallic compound Nb_3Ge has the highest value of critical temperature with $T_C = 23.2$ K.

Following his discovery that electric current can be transported through a superconductor without electrical resistance, Kamerlingh Onnes soon considered technical plans to utilize the phenomenon of superconductivity in cables for the distribution and delivery of electrical power. However, to his great disappointment, during his first experiments he found that the superconducting property is reduced in a magnetic field, and that it completely disappears above a distinct value of the

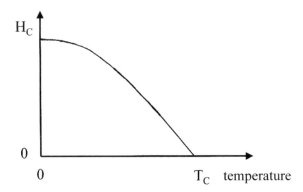

Fig. 8.3 Temperature dependence of the critical magnetic field H_C (schematically)

magnetic field, the "critical magnetic field" H_C. Here an external magnetic field acts in exactly the same way as the "self-field", which is generated by the transported electric current in the superconductor itself. The critical magnetic field $H_C(T)$ vanishes at the critical temperature T_C and increases with decreasing temperature below T_C. It reaches its maximum value at a temperature of zero Kelvin (Fig. 8.3). In many classical superconductors this maximum value ranges between the 100-fold value up to the 5000-fold value of the earth's magnetic field.

Because of the self-field of the transported electric current, the maximum current value, up to which superconductivity is maintained, is limited. This maximum current in a superconductor is referred to as the critical current I_C. In the simplest case the critical current is reached when the magnetic self-field of the current is equal to the critical field H_C. This relationship is also called Silsbee's rule, named after Francis Briggs Silsbee. For many years, this severe restriction on the possibility of transporting electric currents in superconductors has hindered the technical application of superconductivity. This only changed in the 1960s, when new superconducting materials were found with more favorable properties and relatively high values of the critical magnetic field and the critical current. We will come back to this subject at the end of this chapter.

8.1   Meissner Effect, Magnetic Penetration Depth, London Theory

In the year 1933, Walther Meissner (Fig. 8.4) and his collaborator Robert Ochsenfeld at the German Bureau of Standards (*Physikalisch-Technische Reichsanstalt*) in Berlin-Charlottenburg made an important discovery, which turned out to affect strongly subsequent development. If a superconductor is placed within a magnetic field, during the transition to the superconducting state the magnetic field is expelled from the superconductor and vanishes in its interior. Now this phenomenon is referred to as the Meissner effect (Fig. 8.5). At that time after the pioneering

Fig. 8.4 Walther Meissner (*Photo* Physikalisch-Technische Bundesanstalt, Braunschweig and Berlin)

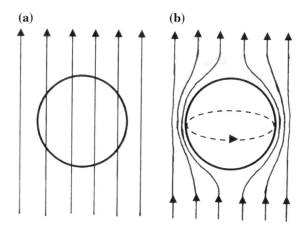

Fig. 8.5 Meissner effect. **a** In the normal state above its critical temperature, the superconducting sphere is completely penetrated by the external magnetic field. **b** Below its critical temperature the superconductor completely expels the magnetic field from its interior as long as the critical magnetic field is not exceeded. The field expulsion is accomplished by means of electric currents, flowing without losses on the surface around the superconductor and thereby shielding the interior of the superconductor against the magnetic field

achievement by Kamerlingh Onnes in Leiden, Walther Meissner was one of the first who could also liquefy the noble gas helium and who managed a properly equipped low-temperature laboratory. After all, it had taken 15 years until outside Leiden at another location, namely by John C. McLennan at the University of Toronto, the

Fig. 8.6 Independence of the
superconducting state of the
path along which this state
has been reached. Because of
the Meissner effect, at point c
the final state with B = 0 is
reached along both paths
a → d → c and a → b → c

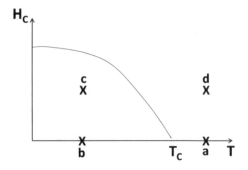

liquefaction of helium had been achieved. Then the Low-Temperature Laboratory of
the German Bureau of Standards was the third location worldwide.

Soon after the Meissner effect was discovered, 1934 Cornelis Jacobus Gorter and
Hendrik Brugt Gerhard Casimir in Holland derived from it an important conclusion.
The magnetic-field expulsion from the interior of the superconductor due to the
Meissner effect, indicates that the superconducting state represents a thermodynamic equilibrium state, which, per definition, is independent of the path along
which this state has been reached by variation of the magnetic field and the temperature. Ultimately it suffices if the temperature T is smaller than the critical
temperature T_C and if the magnetic field is smaller than the critical field $H_C(T)$.

In Fig. 8.6 we have marked point c of the superconducting state (below the
critical temperature T_C and the critical magnetic field H_C). If we assume only
infinite electrical conductivity without the existence of the Meissner effect, along
the path a → b → c the state with B = 0 will be established. On the other hand, the
path a → d → c results in the state of point d with B ≠ 0. Independent of the
previous path, the state with B = 0 is always established only because of the
Meissner effect. (Here we have assumed perfect reversibility of the superconductor,
and have ignored the trapping of magnetic flux due to pinning effects). We rec-
ognize that superconductors are more than perfect electrical conductors (with
infinite electrical conductivity). It is the Meissner effect, which uniquely characterizes the superconducting state.

Furthermore, Gorter and Casimir have shown that the validity of the Meissner
effect yields the possibility of calculating exactly the energy difference between the
normal (nonsuperconducting) and the superconducting state. In their thermodynamic
treatment of the superconducting phase transition, following the discovery of the
Meissner effect, they consider the density of the Gibbs free energy in the normal (G_n)
and the superconducting (G_s) state. In the presence of the magnetic field H one finds:

$$G_s(T, H) = G_s(T, 0) - \int_0^H M(H)\,dH. \qquad (8.1)$$

M(H) is the magnetization. In the case of the Meissner effect (perfect diamagnetism) we have

$$M(H) = -\frac{1}{4\pi}H. \tag{8.2}$$

The last part in (8.1) represents the work performed during the expulsion of the
magnetic field. With (8.2) we obtain

$$G_s(T, H) = G_s(T, 0) + \frac{1}{8\pi}H^2. \tag{8.3}$$

At H = $H_C(T)$ under equilibrium we have $G_n(T, H_C) = G_s(T, H_C)$ and on the
other hand $G_n(T, H_C) = G_n(T, 0)$. Hence, in the case H = H_C for the difference
between the energy density in the normal and the superconducting state we find

$$G_n(T, 0) - G_s(T, 0) = \frac{1}{8\pi}H_C^2(T). \tag{8.4}$$

In order to maintain the Meissner effect, electric currents must flow along the
surface of the superconductor, generating a magnetic field similar to that in an
electric coil. This generated magnetic field is directed opposite to the external
magnetic field and compensates the latter field exactly. These "shielding currents"
must flow along the surface without losses, i.e., without any electrical resistance,
since otherwise the superconducting state cannot last arbitrarily long in the presence
of the magnetic field (Fig. 8.7). The case of electric shielding currents experiencing
losses can be found in each nonsuperconducting electrical conductor such as, for
example, copper. If such a conductor is placed suddenly in a magnetic field, at the
beginning electric shielding currents flow again along its surface, which expel the
magnetic field from the interior of the conductor. However, because of the electric
losses appearing in this case, the shielding currents decrease as a function of time,
and gradually the magnetic field completely penetrates into the electrical conductor.
The time it takes for this decay process of the shielding currents depends on the
electrical conductivity of the conductor. It becomes longer and longer, as the
electrical conductivity increases.

From our discussion it is clearly apparent, that the Meissner effect is based on the
flow of superconducting shielding currents. Hence, superconductivity is the necessary consequence of the existence of the Meissner effect. However, the inverse
conclusion, i.e., that in a material with vanishing electrical resistance the Meissner
effect must exist, is not possible. Therefore, the Meissner effect is more fundamental
for superconductivity than the disappearance of the electrical resistance. However,
the notation "superconductivity" puts the latter quality more into focus. At the time,
Max von Laue looked at the discovery of the Meissner effect as a turning point in
the history of superconductivity.

The superconducting shielding currents near the surface cannot have an arbitrarily high or even infinite density of the electric current flow. Instead, they must

Fig. 8.7 Experimental demonstration of the Meissner effect. A small rectangular piece of a high-temperature superconductor cooled down to the temperature of liquid nitrogen is suspended above a ferromagnetic disk.
A repulsive force exists between this ferromagnet and the shielding currents induced in the superconductor because of the Meissner effect (*Photo* Rainer Straub)

remain limited to a finite value of the current density. This has the consequence, that the shielding currents always need a layer of a specific thickness near the surface, and that the magnetic field penetrates a small but finite distance into the super-conductor, in spite of the existence of the Meissner effect. The thickness of this layer is referred to as the "magnetic penetration depth". In the following we denote this thickness with the symbol λ_m.

The magnetic field of the superconducting shielding current exactly compensates the external magnetic field. The density j_s of the shielding current is approximately

$$j_s = H_C/\lambda_m. \tag{8.5}$$

In 1935, a phenomenological theory of the finite magnetic penetration depth λ_m was proposed by the brothers Fritz and Heinz London. For a short description of their theory we start with the equation of the forces acting upon an electron, without including a dissipative part:

$$m\frac{\partial \mathbf{v_s}}{\partial t} = (-e)\mathbf{E} \tag{8.6}$$

With the superconducting current density

$$\mathbf{j_s} = (-e)\, n_s\, \mathbf{v_s} \tag{8.7}$$

we obtain

$$\mathbf{E} = \left[m/(e^2 n_s) \right] \frac{\partial \mathbf{j_s}}{\partial t} = \mu_o \lambda_m{}^2 \frac{\partial \mathbf{j_s}}{\partial t}. \tag{8.8}$$

The quantity λ_m introduced in (8.5) is also referred to as London penetration depth (and is often denoted by λ_L). The length λ_m is given by

$$\lambda_m{}^2 = m/(\mu_o\, n_s\, e^2). \tag{8.9}$$

Here n_s is the density and $\mathbf{v_s}$ the velocity of the superconducting electrons. μ_o is the permeability of vacuum. With Maxwell's equation

$$\mathrm{curl}\, \mathbf{E} = -\frac{\partial \mathbf{B}}{\partial t} \tag{8.10}$$

from (8.8) we obtain

$$\mu_o \lambda_m{}^2\, \mathrm{curl}\left(\frac{\partial \mathbf{j_s}}{\partial t} \right) + \frac{\partial \mathbf{B}}{\partial t} = 0. \tag{8.11}$$

It was the central idea of Fritz and Heinz London, to extend (8.11) by eliminating the time-derivative and thereby to postulate a new equation

$$\mu_o \lambda_m{}^2\, \mathrm{curl}\, \mathbf{j_s} + \mathbf{B} = 0. \tag{8.12}$$

With Maxwell's equation

$$\mathrm{curl}\, \mathbf{H} = \mathbf{j} \tag{8.13}$$

we obtain

$$\Delta \mathbf{H} = \frac{1}{\lambda_m{}^2} \mathbf{H} \tag{8.14}$$

and from this the solution

$$\mathbf{H}(x) = \mathbf{H}(0)\, \exp(-x/\lambda_m). \tag{8.15}$$

This situation is shown in Fig. 8.8, where we have assumed the geometry of a superconductor with its (positive) x-coordinate running from the surface at x = 0 *to the left* into the interior of the superconductor, which occupies the half-space x > 0. The magnetic field **H** is assumed to be oriented perpendicular to the x-direction.

Equations (8.8) and (8.12) are referred to as first and second London equation, respectively. In addition to Maxwell's equations, they are valid in the case of superconductors and characterize these in contrast to other materials. From (8.15)

Fig. 8.8 Dependence of the density of the super conducting electrons, n_s, and of the magnetic field, H, upon the distance from the interface between a normal (N) and a superconducting (S) region

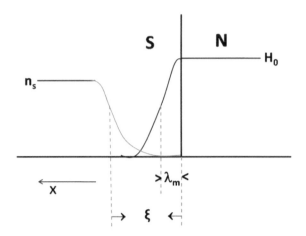

we see, that the magnetic field is shielded exponentially from the interior of the superconductor, and that the shielding occurs within a surface layer of thickness λ_m. At $T \rightarrow T_C$ we find $n_s \rightarrow 0$, and, hence, $\lambda_m \rightarrow \infty$.

For many superconductors the magnetic penetration depth covers the range $\lambda_m = 40\text{--}60$ nm. It strongly increases upon approaching the critical temperature T_C. The magnetic penetration depth represents an important length, which is specific for each material. It plays an important role in many properties of superconductors. For example, because of the finite magnetic penetration depth, an accumulation of small superconducting grains, with the diameter of each grain being similar to the magnetic penetration depth, altogether only displays a strongly reduced Meissner effect, since the magnetically shielded volume fraction remaining in each grain is correspondingly reduced to a relatively small value.

A superconducting circular current, similar to that flowing as a shielding current and causing the Meissner effect, can also serve to find out in a simple experiment, if the electrical resistance in a superconductor is exactly zero, or if a finite residual resistance still remains. For this purpose the usual resistance measurement, based on the electric voltage drop along a current-carrying conductor, is not sufficient, since the electric voltage can be too small to be detected by this method. However, instead of this conventional resistance measurement, one can also start a circular electric current in a superconducting ring by magnetic induction. As in an electric coil, the circular current then generates a magnetic field, that only remains to be detected. Now the task consists in observation of how long the magnetic field of the circular current can be detected. The longer the running time of the current, during which no reduction of the magnetic field is observed, the closer the electrical resistance of the superconducting ring must approach zero. In 1961 the two Americans D. J. Quinn and W. B. Ittner performed an advanced version of such an experiment. By means of two sequentially deposited layers of lead they produced a thin superconducting tube of lead, and they then investigated the temporal decay of the magnetic flux trapped within the tube over a time of seven hours. From their measurements at a temperature of 4 K, as the upper limit of the electrical resistivity of superconducting lead, they obtained the value 3.6×10^{-23} Ω cm. This value is about 17 powers of ten smaller than the resistivity of one of our best metallic conductors, copper, at room temperature.

In addition to the magnetic penetration depth, a second characteristic length plays a fundamental role in superconductors: the "coherence length" ξ. This length indicates the smallest possible spatial distance, within which the property of superconductivity can vary appreciably. In the year 1950, the Englishman Alfred Brian Pippard was the first to point out this spatial rigidity of superconductivity. Also in 1950, the two Russians Vitaly Lazarevich Ginzburg and Lew Dawidowitsch Landau developed another theoretical approach dealing with the question of the spatial coherence of superconductivity. The "Ginzburg-Landau theory" starts from an ansatz for the thermodynamic energy, in combination with the general concept of Landau of "higher-order phase transitions", which are classified according to a specific mathematical scheme. The superconducting property is expressed in terms of a wave function ψ.

Initially, it was felt that between the two characteristic lengths λ_m and ξ the coherence length ξ is always larger than the magnetic penetration depth λ_m. This resulted from the following considerations. Because of the finite extension of the coherence length ξ, a superconducting region cannot exist exactly up to the interface, which separates it from a normal region. Instead, it loses its superconducting property—and, hence, also its superconducting condensation energy—already at the distance ξ from this interface. This results in the positive interface energy $\alpha_1 = (H_C^2/8\pi)\,\xi$. However, from this we have to subtract the amount $(H_C^2/8\pi)\,\lambda_m$, since within the magnetic penetration depth λ_m no gain and, hence, no loss of condensation energy appears. Therefore, the wall energy α of an interface between a normal and a superconducting region is given by

$$\alpha = (H_C^2/8\pi)(\xi - \lambda_m). \tag{8.16}$$

This role of the two lengths ξ and λ_m is shown in Fig. 8.8.

In its simplest form the Meissner effect is observed only if, in the nearest environment of the superconductor, the magnetic field practically remains unchanged during the field expulsion. We have such a case, if the shape of the superconductor is thin and long, and if its longitudinal direction is oriented parallel to the magnetic field. In the other case, if, for example, the superconductor is shaped in the form of a thin plate which is placed perpendicularly within the magnetic field, near the outer border of the plate, the magnetic field is strongly enhanced because of the field expulsion, and it can quickly become larger than the critical magnetic field $H_C(T)$. Now the complete expulsion of the magnetic field cannot be maintained, and magnetic flux will penetrate into the superconductor. As Landau proposed for the first time in 1937, as a consequence, a new state is formed in which both normal domains carrying the local

(d)   **(j)**

(e)   **(k)**

(f)   **(l)**

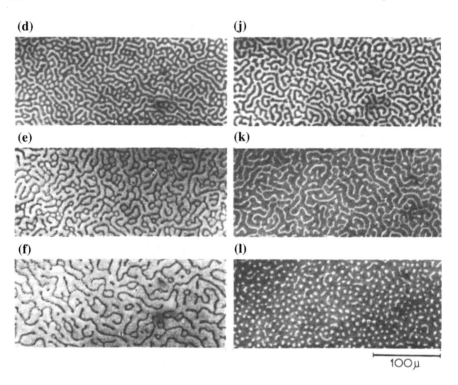

100μ

Fig. 8.9 Intermediate state of a superconducting lead film of 9.3 μm thickness at 4.2 K in increasing (**d** – **f**) and decreasing (**j** – **l**) perpendicular magnetic field at the following values of the magnetc field: **d** 218 G; **e** 348 G; **f** 409 G; **j** 260 G; **k** 101 G; **l** 79 G. Normal domains are bright, superconducting domains are dark

magnetic field H_C and superconducting domains with zero local magnetic field, exist next to each other. This new state is referred to as the "intermediate state". Similar to all spatial systems of domains, the interface separating a normal from a superconducting domain is associated with a specific wall energy. As we see from (8.16), this wall energy is proportional to the length difference $\xi - \lambda_m$. Since initially one had expected, that the wall energy is always positive, and that the formation of a domain wall always consumes energy, one had concluded that the coherence length ξ must be larger than the magnetic penetration depth λ_m.

In Fig. 8.9 we show the intermediate state of a superconducting lead film with 9.3 μm thickness at 4.2 K for different values of the magnetic field oriented perpendicular to the film. The images were obtained magneto-optically by means of a polarization microscope. The normal domains are bright, and the superconducting domains are dark. The critical magnetic field of lead at 4.2 K is 550 G ($T_C = 7.2$ K).

8.2   Type-II Superconductors

However, eventually this picture was shaken. Already in the 1930s the first perturbing signals came from the Low Temperature Laboratory of Leo Vasilyevich Shubnikov in Kharkov in the Ukraine, where experiments on superconductivity had been started at an early stage. Again and again experiments in particular with superconducting alloys, yielded results which could be explained only with great difficulties in terms of the existing ideas. In 1953 the decisive breakthrough was achieved by the young theoretical physicist Alexei A. Abrikosov in Moscow. At the University he was a roommate of Nikolay Zavaritzkii, who performed experiments with superconducting thin films at the famous Kapitza Institute for Physics Problems in order to check the predictions of the Ginzburg-Landau theory. Up to this time one was only interested in the case where the length difference $\xi - \lambda_m$ and, hence, the wall energy, is positive. Now for the first time Abrikosov and Zavaritzkii seriously discussed the possibility, that the length difference could also become negative, if the coherence length ξ were smaller than the magnetic penetration depth λ_m. Based on the Ginzburg-Landau theory, Abrikosov calculated the critical magnetic field for the case, where the difference $\xi - \lambda_m$ is negative, and he could demonstrate, that only in this case could good agreement with Zavaritzkii's experimental data, obtained with particularly carefully prepared thin films, be achieved. Hence, they were apparently dealing with a still unknown, new kind of superconductor. Abrikosov and Zavaritzkii called them the "second group". Eventually, they were referred to as type-II superconductors (with $\xi < \lambda_m$), whereas the superconductors with positive wall energy are now called type-I superconductors (with $\xi > \lambda_m$).

Subsequently, Abrikosov has theoretically analyzed the type-II superconductors in more detail using the Ginzburg-Landau theory and found that, in a magnetic field, they can assume a new state, in which the superconductor is intersected by a regular lattice consisting of individual "magnetic flux quanta". The famous Abrikosov flux-line lattice had been discovered. The state of the superconductor containing the flux-line lattice is referred to as the mixed state (Fig. 8.10). Associated with each magnetic flux line, a spatially confined, local magnetic field passes like a thread through the superconductor. This spatially highly confined magnetic field is generated, as in a magnetic coil, by superconducting circular currents flowing around the thread of the local magnetic field. We will return to this magnetic flux line further below. Abrikosov completed this work in the year 1953. However, the proposed ideas were so novel that they were not accepted by Lew Dawidowitsch Landau, who was Abrikosov's thesis advisor. However, two years later, similar issues appeared in the turbulent flow of superfluid helium at low temperatures. In this case the circulation of the flow is also subject to quantum conditions similar to those of the circular supercurrents associated with the magnetic flux quanta. Only after the American Richard Phillips Feynman had theoretically discussed quantized vortex lines in rotating superfluid helium in this context, was Landau satisfied. In this way it happened that Abrikosov's paper was published only in 1957.

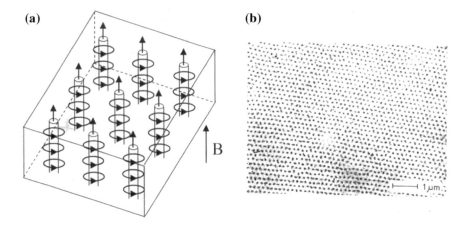

Fig. 8.10 Superconducting mixed state characterized by a lattice of quantized magnetic flux lines, proposed for the first time by Abrikosov. **a** Schematics. A total of nine magnetic flux lines are shown. Each flux line (like a thread carrying a magnetic field) is surrounded by superconducting circular currents. **b** Experimental demonstration, by means of the Bitter decoration technique, of the Abrikosov lattice of magnetic flux lines in a plate of superconducting niobium with 0.5 mm thickness. The many dark spots mark the locations at which the individual magnetic flux lines reach the surface of the superconducting plate (U. Essmann)

In his lecture in Stockholm on December 8, 2003, on the occasion of receiving the 2003 Nobel Prize in Physics, together with Vitaly L. Ginzburg and Anthony J. Leggett, Abrikosov recalled these developments:

> I made my derivation of the vortex lattice in 1953 but the publication was postponed since Landau first disagreed with the whole idea. Only after R. Feynman published his paper on vortices in superfluid helium, and Landau accepted the idea of vortices, did he agree with my derivation, and I published my paper in 1957. Even then it did not attract attention, in spite of an English translation, and only after the discovery in the beginning of the sixties of superconducting alloys and compounds with high critical magnetic fields, did there appear an interest in my work. Nevertheless, even after that the experimentalists did not believe in the possibility of (the) existence of a vortex lattice incommensurable with the crystalline lattice. Only after the vortex lattice was observed experimentally, first by neutron diffraction and then by (Bitter) decoration, did they have no more doubts. Now there exist many different ways to get images of the vortex lattice.

Being asked why Abrikosov did not push more strongly for his spectacular novel results at the time, he gave the following answer:

> The true reason why at the time I did not insist more strongly in my theory, arose from the fact that then all this did not appear so important. Superconductivity was still being considered an exotic phenomenon far from any practical applications. Furthermore, I was already occupied with the extension of quantum electrodynamics to high energies, which appeared to me much more important.

Magnetic flux quanta only penetrate into the interior of a type-II superconductor, when the "lower critical magnetic field" H_{C1} is reached. Below H_{C1} the Meissner effect still exists, and the magnetic field vanishes within the interior of the super-conductor. The mixed state is established above H_{C1} up to the "upper critical magnetic field" H_{C2}. A convincing first experimental confirmation of the existence of the Abrikosov flux-line lattice in the mixed state has been given by Uwe Essmann and Hermann Träuble from the Max Planck Institute for Metals Research in Stuttgart in the year 1967. They succeeded in the imaging of the flux-line lattice at the surface of the superconductor by sprinkling a powder of small ferromagnetic particles onto the surface. Since the powder is attracted by the locations where the flux lines reach the surface, the powder accumulates at these locations forming small piles which decorate the individual flux lines (Fig. 8.10). This decoration method was used for the first time in the year 1931 by the American Francis Bitter for imaging the domain structure of ferromagnetic materials, and since that time it has been referred to as the Bitter technique.

8.3   Magnetic Flux Quantum

Because of the prediction of type-II superconductors and of the magnetic flux-line lattice by Abrikosov, the Ginzburg-Landau theory has achieved great success. By describing the superconducting state of the electrons in terms of a macroscopic quantum mechanical wave function, this theory provided a simple explanation of a series of fundamental phenomena in the field of superconductivity. The magnetic flux quantization is an important example. Within a superconductor magnetic flux can exist only in integer multiples of a smallest unit $h/2e = 2.068 \times 10^{-15}$ Vs, representing the magnetic flux quantum. The quantity h is Planck's constant, and e is the charge of an electron. This quantum condition immediately results from the fact, that the macroscopic wave function describing the superconducting state must reproduce itself exactly, if the spatial coordinate point of the wave function is moved once around the enclosed magnetic flux region and is returned exactly to the starting point. As the smallest unit of magnetic flux, the flux quantum is very tiny. For example, in the magnetic field of the earth one square centimeter is intersected by about one million flux quanta. In beautiful experiments in the year 1961 the two Germans Robert Doll and Martin Näbauer and independently also the Americans Bascom Deaver and William Fairbank demonstrated the quantization of the mag-netic flux in a superconductor. By placing a tiny superconducting tube of only about 10 μm in diameter in a small magnetic field oriented parallel to the axis of the tube, Doll and Näbauer were able to show that the magnetic flux within the small hollow cylinder was either zero or amounted to an integer multiple of the flux quantum specified above (Fig. 8.11).

A more detailed explanation of the step structure shown in Fig. 8.11b is illus-trated in Fig. 8.12. Here in part (a) the superconducting shielding current I_s is plotted versus the magnetic flux density B_e. B_e is oriented parallel to the axis of the small superconducting cylinder. The magnetic flux $\pi R^2 B_e$ (R = cylinder radius) passing through the cross-sectional area of the cylinder is indicated in units of the

Fig. 8.11 Experimental proof of the magnetic flux quantization in a superconductor. **a** A tiny superconducting tube with only about 10 μm diameter is cooled down in the presence of a small magnetic field oriented parallel to the axis of the tube. Below the critical temperature T_C the magnetic field is turned off, and the magnetic flux trapped within the tube is measured. **b** The frozen-in magnetic flux displays a quantized step structure as a function of the magnetic field B, since only integer multiples of the magnetic flux quantum (h/2e) are allowed within the tube. The Figure shows the observation of 0, 1, and 2 magnetic flux quanta, respectively. Without magnetic flux quantization, the data points would fall on the straight dashed line (R. Doll and M. Näbauer)

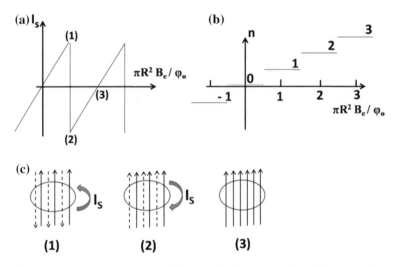

Fig. 8.12 Experimental demonstration of the magnetic flux quantization during entry of magnetic flux into a small superconducting cylinder. **a** Superconducting shielding current I_s plotted versus the magnetic flux density B_e oriented parallel to the cylinder axis. **b** Number n of the magnetic flux quanta within the cylinder plotted versus B_e. **c** Superposition of the applied magnetic field (solid arrows) and of the magnetic field generated by I_s (dashed arrows) at the three points 1–3 from (**a**). Further details are given in the text

magnetic flux quantum $\varphi_o = h/2e$. (The vector $\boldsymbol{\varphi}_o$ is oriented parallel to the flux density $\mathbf{B}$). Initially, the shielding current I_s prevents the entry of magnetic flux into the cylinder due to the Meissner effect. When the magnetic flux density has reached the value $B_e = \varphi_o/(2\pi R^2)$, the shielding current compensates exactly half a flux quantum $\varphi_o/2$ within the cylinder (point 1). If B_e is increased further, the shielding current I_s reverses its sign (instead of increasing further), such that exactly one flux quantum φ_o exists within the cylinder (half of which is generated by I_s; point 2). When B_e continues to be increased, $|I_s|$ decreases again until (at point 3) the state with $I_s = 0$ is reached at $B_e = \varphi_o/(\pi R^2)$. During further increasing B_e, this process repeats itself. In this way, the steps of the number n of the magnetic flux quanta within the cylinder are generated (Figs. 8.11b and 8.12b). In Fig. 8.12c the superposition of the applied magnetic field (solid arrows) and of the magnetic field generated by I_s (dashed arrows) is shown schematically for the three points 1–3 of Fig. 8.12a. The entry of the magnetic flux quanta φ_o into the cylinder effects, that the shielding current I_s and the kinetic energy associated with it remain limited.

Theoretically, the magnetic flux quantization is demonstrated as follows. Representing the superconducting state by a macroscopic wave function (order parameter)

$$\psi(\mathbf{r}, t) = |\psi(\mathbf{r}, t)| e^{i\varphi(\mathbf{r},t)} \tag{8.17}$$

with an amplitude $|\psi(\mathbf{r}, t)|$ and a phase $\varphi(\mathbf{r}, t)$, then the exact reproduction of the state after a complete path around a fixed point in the superconductor requires that the ring integral along this path is

$$\oint \mathbf{grad}\, \varphi \cdot d\mathbf{s} = n\, 2\pi \tag{8.18}$$

where n is an integer. In a magnetic field, the generalized particle momentum is

$$m^* \mathbf{v}_s = \hbar\, \mathbf{grad}\, \varphi - e^* \mathbf{A} \tag{8.19}$$

which yields

$$\oint (m^* \mathbf{v}_s + e^* \mathbf{A})\, d\mathbf{s} = n\, h \tag{8.20}$$

Here $\mathbf{A}$ is the vector potential and $d\mathbf{s}$ a line element. Assuming that the superconductor is sufficiently thick, we can find an integration path where $\mathbf{v}_s = 0$. Then applying Stokes' law, we have

$$\oint \mathbf{A}\, d\mathbf{s} = \int \mathrm{curl}\, \mathbf{A}\, d\mathbf{a} = \int \mathbf{B}\, d\mathbf{a} \tag{8.21}$$

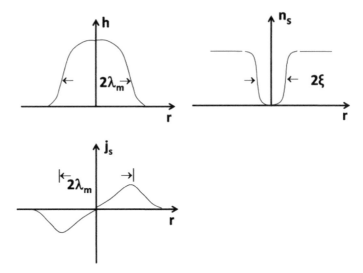

Fig. 8.13 Structure of a single flux line. Local magnetic field h, density of the superconducting electrons n_s, and the circulating superconducting current density j_s as a function of the distance r from the axis of the flux line

(**da** is an area element), and with $e^* = 2e$ because of Cooper pairing, (8.20) yields the magnetic flux quantization: $\int \mathbf{B} \, \mathbf{da} = n\frac{h}{2e} = n\,\varphi_0$.

Now we will look more closely at a magnetic flux line in a type-II superconductor, discussed for the first time by Abrikosov (Fig. 8.13). In its center each flux line has a normal core, the radius of which is approximately given by the coherence length ξ. The local magnetic field associated with the flux line reaches a maximum in the center of the line and decreases toward the outside. This decrease in the magnetic field mostly takes place within a radius given by the magnetic penetration depth λ_m. This spatial confinement of the local magnetic field is accomplished by means of circulating superconducting currents flowing around the center of the flux line within a radial distance, in the range between the coherence length ξ and the magnetic penetration depth λ_m.

The expression (h/2e), indicated above for the magnetic flux quantum in superconductors, is exactly half of the value (h/e) of the magnetic flux quantum, which we have dealt with in Chap. 7 in the context of the fractional quantum-Hall-effect. The reason for this half-value is the fact that superconductivity is based on the Cooper pairs, which consist of two electrons. This subject will be taken up next.

8.4 BCS Theory, Energy Gap

It has taken nearly 50 years since the discovery of superconductivity, until for the first time a microscopic theory was proposed which could explain satisfactorily the underlying mechanism. In the year 1957 the three Americans John Bardeen, Leon Cooper, and Robert Schrieffer achieved the long-expected theoretical breakthrough. Their theory, the "BCS theory", quickly became very famous. The question why it took so long to produce a theoretical explanation of superconductivity, can be answered relatively simply. The energy difference of the electrons between their normal and their superconducting state is extremely small and much smaller than the Fermi energy. On the other hand, the uncertainty in the calculation of the different individual contributions to the energy of the electrons in the crystal is much larger than the energy gain during the transition into the superconducting state. Hence, the theory had to find the exact point leading to superconductivity. The BCS theory is based on the central idea that, at low temperatures, a specific attractive force is acting between two electrons. Because of this attraction, two electrons combine into pairs in a distinctive way and experience an energy reduction in the form of binding energy. Such a formation of pairs accompanied by a reduction in energy had been theoretically derived by Leon Cooper in 1956. Therefore, the electron pairs are referred to as "Cooper pairs". According to the BCS theory, the attractive force leading to the formation of the Cooper pairs is due to the distortions of the crystal lattice near the individual electrons, i.e., due to the phonons. In this way, the otherwise expected repulsive force between two electrons is overcompensated. Already by the early 1950s strong indications for the important role of the crystal lattice in the mechanism of superconductivity were obtained, based on experimental observations of the "isotope effect". One generally speaks of an isotope effect, when the result depends on the mass of the atomic nuclei at constant electric charge of the nuclei, i.e., on the number of neutrons in the atomic nuclei. By careful study of the different and specially prepared pure isotopes of various superconducting materials such as, for example, lead, mercury, and tin, it was found, that the critical temperature T_C is inversely proportional to the square-root of the mass of the lattice atoms:

$$T_C \sim 1/M^{\alpha} \tag{8.22}$$

with the exponent $\alpha = 0.5$. Hence, the crystal lattice must play some role in superconductivity.

During pair formation, two electrons with opposite spin always combine with each other. Therefore, the total spin of an individual Cooper pair is zero, and the Pauli principle does not apply in this case. Hence, all Cooper pairs can occupy the same quantum state, which is described in terms of a macroscopic quantum mechanical wave function. However, not all electrons participate in the formation of Cooper pairs and in the macroscopic quantum state. Instead, only the electrons from a distinct small energy interval near the Fermi surface are involved. We see

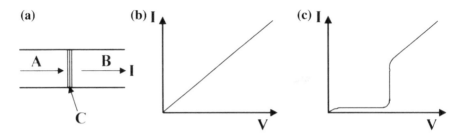

Fig. 8.14 Experimental proof of the energy gap in a superconductor by means of the tunneling experiment of Giaever. **a** A superconducting electrode A and a normal electrode B are separated from each other by a thin, electrically insulating barrier C, such that the electric current flow across the barrier is only possible because of the quantum mechanical tunneling process. **b** Electric current I plotted versus the voltage V in the case when both electrodes are metals in the normal state. **c** Electric current I plotted versus the voltage V in the case when one metal electrode is superconducting. The electric current can start to flow only when the electrical potential difference between the two electrodes has reached the value of the energy gap

again, how the concept of the Fermi surface plays a central role. Mathematically, the subject of superconductivity confronts us with a "manybody problem", requiring special techniques for its theoretical treatment. The development of these necessary new methods started about 60–70 years ago in conjunction with quantum field theory. The first steps of this theory can be found in a paper published in the year 1928 by the German Pascual Jordan and Eugene Paul Wigner from Hungary.

One main result of the BCS theory was the prediction that, in the superconducting state, a gap appears in the energy spectrum of the electrons at the Fermi energy, in which no energy states exist which can be occupied by electrons. The energy gap vanishes above the critical temperature T_C. Below T_C the energy gap increases with decreasing temperature in a distinct way and reaches its maximum value at a temperature of zero Kelvin. In the year 1960, Ivar Giaever presented an impressive proof of this energy gap by means of his famous tunneling experiment (Fig. 8.14). For some time he had been fascinated by the quantum-mechanical tunneling effect. Giaever was born in Norway and as a young mechanical engineer was employed at General Electric in Schenectady in the American Federal State of New York. At the Rensselaer Polytechnic Institute near the location of his employment, he had heard in a lecture about the new BCS theory and its prediction of a gap in the energy spectrum of the electrons. On his way home after the lecture he had the idea that the energy gap must directly affect the electric current flow between a superconducting and a normal electrode, if the two electrodes are separated from each other by a thin, electrically insulating barrier. Because of this barrier, the electric current flow is possible only by means of the quantum mechanical tunneling process. Hence, this arrangement is referred to as a tunnel junction. During the propagation of particles, the tunneling effect is caused by the fact, that the wave function of the particle can still seep through a high wall and can reach an appreciable value at the other side. However, in our tunnel junction the tunneling current cannot flow as long as no allowed energy states in the super-conductor are available for the electrons coming from the other electrode, because of the energy gap. Only when the potential difference between both electrodes has reached the value of the energy gap because of the applied electric voltage does the electric current flow become possible. We have a similar result, if both electrodes are superconducting. Hence, it should be possible to determine the superconducting energy gap just by means of a simple measurement of the electric voltage and the electric current in a tunnel junction. Giaever's experiments have impressively confirmed these expectations. After this pioneering step, tunneling experiments with superconductors have become an important source of information about the physics of superconductors. The BCS theory has been confirmed by many further experiments and has quickly found wide acceptance. There exists a long list of physicists, who had tried before without success to construct a microscopic theory of the mechanism of superconductivity. Among others, this list includes the names Felix Bloch, Niels Bohr, Léon Brillouin, Jakov I. Frenkel, Werner Heisenberg, Ralph Kronig, Lew Dawidowitsch Landau, and Wolfgang Pauli.

The fact that it is the formation of Cooper pairs occupying a macroscopic quantum state, which leads to superconductivity, is also visible in the magnitude of the magnetic flux quantum discussed above. Since the Cooper pairs are composed of two elementary charges, the magnetic flux quantum (h/2e) is only half as large as would be the case if the underlying elementary particles carried only a single elementary charge, leading to (h/e).

8.5 Josephson Effect

Soon after Giaever had published the result of his famous tunneling experiment, a student in Cambridge, England was interested in the underlying tunneling process: Brian David Josephson. He was tutored by Alfred Brian Pippard, and in 1961/1962 he attended lectures by the American Philip Warren Anderson about the new developments in the theory of superconductivity. Josephson was highly impressed by the concept of superconductivity in terms of a macroscopic quantum phe-nomenon, which extended far beyond the range of validity in individual atoms or molecules. When he theoretically analyzed the details of the electric current flow through the barrier of a tunnel junction between two superconductors, as it had been used by Giaever, he derived two equations for the electric current and for the electric voltage, respectively, which are known since as the Josephson equations:

$$I_s = I_C \sin \chi \tag{8.23}$$

$$\frac{\partial \chi}{\partial t} = \frac{2e}{\hbar} V \tag{8.24}$$

In (8.23) the current of Cooper pairs flowing without electrical resistance is described. Equation (8.24) indicates that an electric voltage V across the tunnel junction is always accompanied by an alternating supercurrent between both superconductors oscillating at a high frequency. The frequency of these Josephson oscillations increases proportionally with the electric voltage. Equations (8.23) and (8.24) are based on the concept that superconductivity represents a macroscopic quantum phenomenon, described by the wave function (8.17) with an amplitude $|\psi(\mathbf{r}, t)|$ and a phase $\varphi(\mathbf{r}, t)$. The current-phase relation (8.23) indicates that the supercurrent I_s flowing across a weak contact is connected with the phase difference $\chi = \varphi_2 - \varphi_1$ between both sides of the junction. I_C is the critical current of the junction geometry.

The Josephson equations (8.23) and (8.24) can be derived in different ways. One derivation given by Richard Feynman starts with the time-dependent Schrödinger equation for the two wave functions ψ_1 and ψ_2 of the two superconductors, which are still separated initially, and then a coupling between both is added. Denoting the two sides of the Josephson junction by the indices 1 and 2, respectively, the Schrödinger equations of the wave function for each side are

$$i\hbar \frac{d\psi_1}{dt} = \mu_1 \psi_1 + K \psi_2 \tag{8.25}$$

and

$$i\hbar \frac{d\psi_2}{dt} = \mu_2 \psi_2 + K \psi_1 \tag{8.26}$$

Here K is a coupling constant, and $\mu_{1,2}$ the chemical potential (energy) on each side. With the potential difference V between both sides we have

$$\mu_1 = e^*V/2; \; \mu_2 = -e^*V/2 \tag{8.27}$$

Connecting the wave function (8.17) with the density n_s of the superconducting electrons according to $|\psi|^2 = n_s$, we write

$$\psi_1 = n_{s1}^{1/2} \cdot e^{i\varphi_1}; \quad \psi_2 = n_{s2}^{1/2} \cdot e^{i\varphi_2} \tag{8.28}$$

and by insertion into (8.25) we obtain

$$i\hbar \, e^{i\varphi_1} \frac{1}{2n_{s1}^{1/2}} \dot{n}_{s1} - \hbar \, n_{s1}^{1/2} \dot{\varphi}_1 \, e^{i\varphi_1} = \frac{e^*V}{2} n_{s1}^{1/2} \cdot e^{i\varphi_1} + K \, n_{s2}^{1/2} \cdot e^{i\varphi_2} \tag{8.29}$$

Insertion into (8.26) yields an analogous result. In (8.29) we divide by $\hbar \, e^{i\varphi_1}/2 \, n_{s1}^{1/2}$ and find

$$i\,\dot{n}_{s1} - 2\,n_{s1}\,\dot{\phi}_1 = \frac{e^*V}{2\hbar}\,2n_{s1} + \frac{K}{\hbar}\,2(n_{s1}n_{s2})^{1/2}e^{i(\phi_2-\phi_1)} \tag{8.30}$$

Equalizing the *real* parts on both sides yields

$$-2n_{s1}\,\dot{\phi}_1 = \frac{e^*V}{\hbar}\,n_{s1} + \frac{2K}{\hbar}(n_{s1}\,n_{s2})^{1/2}\cos(\phi_2 - \phi_1)$$

and

$$\dot{\phi}_1 = -\frac{e^*V}{2\hbar} - \frac{K}{\hbar}(n_{s2}/n_{s1})^{1/2}\cos(\phi_2 - \phi_1) \tag{8.31}$$

Analogous one finds from (8.26)

$$\dot{\phi}_2 = \frac{e^*V}{2\hbar} - \frac{K}{\hbar}(n_{s1}/n_{s2})^{1/2}\cos(\phi_2 - \phi_1) \tag{8.32}$$

In the case $n_{s1} = n_{s2} = n_s$ we have

$$\dot{\phi}_2 - \dot{\phi}_1 = \frac{e^*V}{\hbar} \tag{8.33}$$

Setting $(\phi_2 - \phi_1) = \chi$ and noting the Cooper pairs, $e^* = 2e$, finally we obtain (8.24).

This last result reminds us of the Einstein relation $2\,eV = \Delta E = \hbar\omega$. In Sect. 8.7 we return to relation (8.24), when we discuss its application in the form of the Josephson voltage standard.

Equalizing the *imaginary* parts in (8.30), we obtain

$$\dot{n}_{s1} = \frac{2K}{\hbar}(n_{s1}\,n_{s2})^{1/2}\sin\chi \tag{8.34}$$

and analogous from (8.26)

$$\dot{n}_{s2} = -\frac{2K}{\hbar}(n_{s1}\,n_{s2})^{1/2}\sin\chi \tag{8.35}$$

With $\dot{n}_{s1} = -\dot{n}_{s2}$ being proportional to the current I_s, we recover the current-phase relation (8.23).

Josephson made both predictions in the year 1962. At first, his theory was met by scepticism and hardly any understanding, as often happens with completely new ideas. As an example, Felix Bloch speaks of a conversation he had with the, also highly renowned, American theoretical physicist Chen Ning Yang:

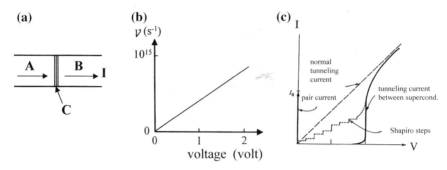

Fig. 8.15 Josephson oscillation of the supercurrent between the superconducting electrodes of a tunnel junction in the presence of an electric voltage across the junction. **a** In a Josephson junction, the two superconducting electrodes A and B are only weakly coupled to each other, for example, by means of a thin, electrically insulating barrier C, which allows electric current to flow only because of the quantum mechanical tunneling process. **b** The frequency ν of the Josephson oscillation of the supercurrent between the two electrodes increases proportionally with the electric voltage V across the junction. At a voltage of one volt the frequency is about 483,000 GHz. **c** Electric current I plotted versus the voltage V of a Josephson junction. The solid and the dashed curve show the tunneling current in the case of superconductivity and in the case of normal conductance, respectively. At zero voltage we see the Josephson pair current flowing without resistance up to its maximum value I_0. During irradiation of the junction with microwaves the current-voltage characteristic displays the "Shapiro steps", which are caused by the combined action of the Josephson oscillation within the junction and the microwaves

> Yang told me that he could not understand it, and asked whether I could. In all honesty I had to confess that I could not either, but we made a deal that whoever of us first understood the effect would explain it to the other.

By 1963 Josephson's theory had already been confirmed experimentally (Fig. 8.15). The second Josephson equation also emphasizes again, that it is the Cooper pairs with their two elementary charges, which lead to superconductivity.

In this context of our discussion of the Josephson effect between two superconductors, only weakly coupled to each other, a brief note in the protocol of the board meeting at the German Bureau of Standards (*Physikalisch-Technische Reichsanstalt*) in Berlin-Charlottenburg in March 1926 is interesting historically. At that time, Albert Einstein was a member of the board, and during the meeting he made the following remark: "*Of particular interest is the question, of whether the location of the junction joining two superconductors also becomes superconducting*".

The remark by Einstein at the board meeting of the *Reichsanstalt* is connected with the fact that, at the time, for explaining superconductivity he had proposed that it is caused by molecular conducting chains. In his commemorative address for Kamerlingh Onnes in 1923, Einstein discussed the state of superconductivity as follows:

So it appears unavoidable that the superconducting currents are carried by closed molecular chains (conducting chains), the electrons of which incessantly experience cyclical exchanges. Therefore, Kamerlingh Onnes compares the closed currents in superconductors with Ampère's molecular currents. […] It may appear unlikely, that different kinds of atoms can combine to form conducting chains. Hence, the transition from one superconducting metal to another is perhaps never superconducting.

Motivated by these ideas of Einstein, in 1932 Walther Meissner together with Ragnar Holm, working for the Siemens company in Berlin in its research laboratories, carried out experiments concerning the contact between two superconductors. They performed the measurements on superconducting tin (Sn) and lead (Pb) and investigated Sn-Sn, Pb-Pb, and Sn-Pb contacts. Meissner and Holm reported their results in an article in *Zeitschrift für Physik* and concluded:

Between superconductors from the same or from different materials a superconducting contact is possible without welding both materials together. During entry of superconductivity the resistance of the contact layer also vanishes.

Hence, the model of the molecular conducting channels, proposed by Einstein, had been disproved.

In this case, motivated by the ideas of Einstein, for the first time the subject of thin contacts or micro-bridges between two superconductors, which subsequently would gain great significance, was brought up. However, it took another 30 years, until Brian David Josephson in 1962 finally answered these questions.

8.6   Motion of the Magnetic Flux Quanta

Again and again, the scientists were concerned with the question whether, in a superconducting ring, the supercurrents really flow forever, or whether they decay perhaps extremely slowly. In the beginning of the 1960s, this question became very urgent, when at the Bell Laboratories Bernd T. Matthias, working together with Theodore Geballe, discovered the new superconducting niobium alloys Nb_3Sn and NbZr. In this context we note that for a long time superconductor research had concentrated primarily on high-purity metals. An early exception of this trend is a brief report by Walther Meissner from 1929 dealing with the observation of superconductivity in a wire of copper-sulfide (CuS) with a critical temperature $T_C = 1.6$ K. Subsequently, this discovery may have led Enrico Fermi at the University of Chicago to urge two young postdocs, Bernd Matthias and John Hulm, to start a wide search for more superconductors.

Returning to Nb_3Sn and NbZr, the tests performed by Gene Kunzler indicated promising high values of the critical electric current density and of the critical magnetic field in these materials. The further experiments, performed in particular by the Korean Young Kim, showed that there exists a "critical state", above which in a superconducting ring the current always decreases. The crucial idea then came from Philip W. Anderson, when he recognized that this process does not set in discontinuously, but is caused by the motion of the magnetic flux quanta. The concepts of *flux creep* and *flux flow* were born.

This highly important consequence of the magnetic flux quanta (in addition to their influence on the magnetic properties) is due to the following. If, under the influence of a force, the flux quanta move within the superconductor, an electric field and, hence, an electric voltage is generated. This "flux-flow voltage" is proportional to the velocity and to the number of the moving flux lines. An electric current of density $\mathbf{j}$ in the superconductor causes the Lorentz force $\mathbf{f}_L = \mathbf{j} \times \boldsymbol{\varphi}_o$ acting on the flux quanta. It is oriented perpendicular to the direction of the electric current and to the magnetic field of the flux lines. The Lorentz force can cause a motion of the magnetic flux lines which, in turn, generates the electric field $\mathbf{E}$

$$\mathbf{E} = -\mathbf{v}_\varphi \times \mathbf{B}. \tag{8.36}$$

$\mathbf{v}_\varphi$ is the velocity of the flux lines. This electric field is always oriented perpendicularly to the motional direction and to the magnetic field of the magnetic flux lines. Hence, in the case of the Lorentz force the electric field and the electric current have the same direction, such that the flux-line motion in the superconductor causes electric losses. The process of the flux-line motion follows the (here simplified) force equation

$$\mathbf{j} \times \boldsymbol{\varphi}_o - \eta \, \mathbf{v}_\varphi = 0. \tag{8.37}$$

Here $\eta \, \mathbf{v}_\varphi$ is the dissipative contribution and η a damping constant. In (8.37) the forces are given per unit length of the flux lines. From (8.36) and (8.37) we obtain the flux-flow resistivity

$$\rho_f = \varphi_o \, B / \eta. \tag{8.38}$$

It is exactly this mechanism, which always limits the electric current flow without electrical resistance and without losses in a superconductor. Therefore, it is of utmost interest, to prevent this process of flux-line motion as much as possible. In this last discussion we have simplified the situation by neglecting in (8.37) the effects of the pinning forces and a force component which causes the Hall-effect during the motion of the flux lines.

In Fig. 8.16 as an example, we show the flux-flow voltage as a function of the electric current in a niobium foil for different values of the magnetic field oriented perpendicular to the foil. Thickness of the foil = 18 μm, width of the foil = 4 mm, T = 4.22 K. The electric resistance ratio R(295 K)/R(4.2 K) was 620, where R(4.2 K) was measured in a perpendicular magnetic field of 4000 G. The voltage sets in at a finite critical current and then shows an upward curvature, after which it increases linearly with increasing current as expected from (8.37) and (8.38). The finite critical current results from the pinning forces due to the spatial inhomogeneity of the superconductor, hindering as pinning centers the motion of the magnetic flux

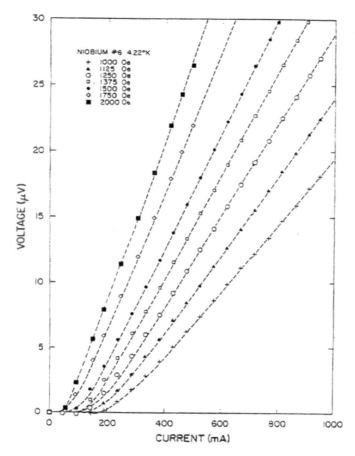

Fig. 8.16 Flux-flow voltage versus the electric current in a niobium foil of 18 μm thickness and 4 mm width for different perpendicular magnetic fields. T = 4.22 K; T_C = 9.2 K

lines. The slope of the linear curve sections represents the flux-flow resistance and increases with increasing external magnetic field.

In Fig. 8.17 we show schematically the specific flux-flow resistance ρ_f plotted versus the external magnetic field H. Initially, ρ_f increases linearly with increasing H, then turning into a much steeper curve section, which near the upper critical magnetic field H_{C2} approaches the normal value ρ_n of the specific resistance.

The motion of a magnetic flux quantum in a superconductor resulting in the flux-flow resistance and in the destruction of superconductivity is an example of a general principle of nature, according to which the generation and the motion of a local defect through an otherwise homogeneous system leads to major macroscopic effects. In other words: here a little cause can achieve a major effect. In Chap. 12 we will discuss a similar example, in which the generation and the motion of individual dislocations through an otherwise homogeneous crystal, affects the mechanical

Fig. 8.17 Schematics of the specific flux-flow resistance ρ_f in a type-II superconductor plotted versus the magnetic field for different temperatures ($T_1 > T_2 > T_3$; ρ_n = normal resistance)

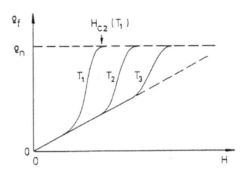

properties and results in the deformation of the crystal at unexpectedly small values of the mechanical tension.

> The motion of local vortices within a supra-fluid also plays a role in the dynamics of the neutron stars or pulsars. These stars rotate having a rotational frequency above 1/sec. However, one observes that for a long time this rotation becomes slower, until abruptly the rotational speed jumps to a higher value. Today this behavior is explained by the motion and abrupt jumps of local vortices.

Material scientists have been taking great pains to pin the magnetic flux lines at specific locations by introducing "pinning centers" into the superconductor. In this way, one hopes that the flux lines are not moving any longer under the influence of the Lorentz force, or that this motion and the electric losses only start to appear at electric currents as high as possible. In recent years great effort has been devoted to this subject in material science and metallurgy. These developments were motivated by the interest in the possibilities for the technical applications of superconductivity. Next we will turn to the technical applications and look at a few examples.

8.7   Technical Applications

For the applications of superconductivity in electronics and microelectronics the magnetic flux quantization and the Josephson effect are of central interest. Both phenomena are intimately connected with the nature of superconductivity as a macroscopic quantum phenomenon and with the description of the state of the Cooper pairs in terms of a quantum mechanical wave function. Here the limitation of the quantum-theoretical concepts to atomic and subatomic objects, is suspended. Instead, these concepts are directly technically utilized in devices and instruments.

An electronic instrument used today in many different ways is the "SQUID" (abbreviated from Superconducting Quantum Interference Device). It is based on the magnetic flux quantization and the Josephson effect. A small closed superconducting loop is interrupted by two Josephson junctions connected in parallel. If the loop is penetrated by magnetic flux, within the loop the magnetic flux can exist only in units of integer multiples of a magnetic flux quantum. This quantum condition is satisfied by means of the induction of a circulating superconducting shielding current within the loop, in such a way that the generated additional magnetic flux in the loop, in combination with the external magnetic flux, exactly supplement each other to yield an integer multiple of a magnetic flux quantum (similar as it is shown in Fig. 8.12c). As a result one observes an exactly periodic modulation of the shielding current within the loop as a function of the external magnetic field, where the length of the magnetic period corresponds exactly to one magnetic flux quantum in the loop. It is important that the circulating shielding current should always be added to an external electric current which is also flowing through the device. As a consequence, the measured electric voltage along the loop configuration with the two parallel Josephson junctions also displays a periodic modulation. Since even a small fraction of a single modulation period of the magnetic field can be resolved during the measurement of the electrical resistance, an extremely high sensitivity of the magnetic field measurement is achieved. Today, the fabrication of SQUIDs is carried out usually by means of thin-film and integrated circuit technology.

As sensors for detecting magnetic fields, SQUIDs have the highest sensitivity which can be reached today. This fact results in many of the applications of SQUIDs, for example, in the field of research or nondestructive material testing. Interesting applications are also found in medical diagnostics for detecting the magnetic fields generated by the electric currents from cardiac activity or in the brain. In this way, the new fields of magneto-cardiography and magneto-encephalography developed only because of extremely sensitive SQUIDs. For example, today instruments with a total of up to 275 SQUID channels are available for brain research, where the channels with the individual sensors are arranged in a three-dimensional way around the head of the test person or of the patient (Fig. 8.18). Very recently, highly miniaturized SQUIDs have been used also in SQUID-scanning microscopes. At an extremely high magnetic-field sensitivity, these instruments achieve a spatial resolution as high as only a few μm, such that individual magnetic flux quanta in a superconductor can be nicely imaged. (In Fig. 9.8 of Chap. 9 we present an application of a SQUID-scanning microscope.)

Presently, small spin systems in nano-particles are moving to the center of attention. The newest development of this SQUID instrumentation concerns the fabrication of ultra-small devices on sharp nano-scale tips (nano-SQUID-on-tip). By depositing superconducting lead or niobium at the apex of hollow quartz tubes, SQUID loops with an effective diameter of only 160 nm or even less than 100 nm could be achieved. It is estimated that the signal of a single electron spin located 10 nm below a SQUID-on-tip loop can be resolved with a spatial resolution of about 20 nm.

We discussed above, that an electric voltage drop at a Josephson junction is always associated with a high-frequency oscillation of the supercurrent flowing between the two electrodes of the junction, as predicted by the second Josephson equation (8.24). Here an electric voltage of 10^{-3} V corresponds to an oscillation frequency of 483.6 GHz (Gigahertz). Vice versa, distinctly sharp electric voltage plateaus result at the current-carrying Josephson junction, if the junction is

Fig. 8.18 Magneto-encephalography. Left: Whole system with the test person carrying the helmet containing the SQUID magnetic-field sensors, within a magnetically shielded chamber. Right: View into the helmet having an arrangement of 151 SQUID sensors (*Photos* MEG International Services Ltd.)

irradiated with a high-frequency electromagnetic wave such as a microwave, for example. Then the magnitude of the voltage plateau is unequivocally fixed by the frequency of the irradiating electromagnetic wave because of the second Josephson equation. This exact quantum condition between a frequency and an electric voltage, in combination with the fact that frequencies can be determined extremely accurately, was the reason why, since January 1, 1990, the legal definition of the electric voltage unit established by the National Bureaus of Standards is based on the Josephson effect in the form of the "Josephson voltage standard". Officially, a voltage of one volt corresponds to a frequency 483597.9 GHz.

In Chap. 7 we discussed another quantum definition of an electric unit, namely the von-Klitzing effect for the definition of the unit of electrical resistance. Together with the Josephson oscillation as the combining mechanism between an electric voltage and a frequency, two sides of the famous quantum triangle consisting of current, voltage, and resistance, for the definition of the electric units, are now completed. The remaining third side, yielding the connection between an electric current and a frequency, is presently the subject of ongoing research in different laboratories. Here the goal is to define the electric current in terms of the frequency of transfer of individual electrons. With the example of the Josephson voltage standard we will conclude our discussion of the field of Josephson electronics and Josephson technology, which today is already well developed for measuring electronics.

For many years the relatively low values of the critical magnetic fields and of the critical currents had prevented the technical high-current applications of superconductivity in energy technology and in electric machinery. This changed only in the 1960s, when new superconducting materials with higher values of the critical electric current density and of the upper critical magnetic field H_{C2} became known. Then the compounds NbTi with $T_C = 9.6$ K and Nb_3Sn with $T_C = 18$ K, technically became highly relevant. Among the classical superconductors, thin layers of the compound Nb_3Ge showed the highest critical temperature with $T_C = 23.2$ K. For the fabrication of wires. special drawing procedures and different mechanical processing stages with an optimized combination of heat treatment and cold-work were developed. Particularly successful were the "multifilamentary wires", where many thin filaments of the superconducting material are imbedded within a copper matrix. This technique ensures that, during the breakdown of superconductivity, because of overloading a certain finite electrical conductivity still remains, and that on the other hand there exists a sufficiently large number of pinning centers in order to pin the magnetic flux quanta in the superconductor.

One of the main applications of technical superconductors can be found in magnetic coils. Today, superconducting magnets are used in large numbers in research laboratories (Fig. 8.19). Particularly large versions serve as beam-guiding magnets of particle accelerators and are also important components of the associated particle-detector systems. Since a few years, the largest superconducting accelerator plant worldwide, the *Large Hadron Collider* (LHC), is operating at the European Nuclear Research Center (CERN) in Geneva. It is placed within a circular tunnel of 27 km length and, at full operation, accelerates protons up to energies of 7000 GeV. For beam guidance, a total of more than 1600 superconducting magnets, constructed from NbTi, are used. Most magnets weigh more than 27 t. For this more than 7000 km of NbTi wire were needed. In addition, more than 3500 superconducting correcting magnets are used. Only the correct adjustment of all magnets allows the storage of highly accurate proton beams, as they are required by the LHC. During operation, a total of 31,000 t material must be cooled down to 1.9 K. This requires 12 million liters of liquid nitrogen and 700,000 liters of liquid helium. (In Chap. 1 we had mentioned the extremely large cryogenic facilities installed at the LHC.)

Another large-scale application of superconducting coils is found in the case of the magnetically levitated train. In particular the Japanese JR-Maglev project recently has seen promising advances. In 2015 during tests, a speed above 600 km/h has been reached. Superconducting coils generating magnetic fields of about 5 T are mounted within the train. Electrically well conducting loops are imbedded along the gliding track, in which strong eddy currents are generated during train motion. According to Lenz' rule, the magnetic field of the eddy currents repels the field of the coils above. In this way the levitating force is generated. Since this repulsion becomes sufficiently strong only above a certain minimum speed, initially the train has to move on wheels, which are retracted when this speed is reached.

During the past 15 years superconducting magnets for medical nuclear spin tomography have developed into the most important market of superconductor technology. This started in the beginning of the 1980s when the health authorities worldwide approved the use of nuclear spin tomography in medical diagnostics. Presently, the annual turnover of industry in this field amounts to 2–3 billion €.

(a)   **(b)**

Fig. 8.19 Superconducting magnetic coils. **a** Commercially available coil for research purposes. The coil is wound from niobium-titanium (NbTi) wire and can generate a magnetic field of up to 9 T, corresponding to about one million times the magnetic field of the earth (Oxford). **b** Superconducting model coil with its test set-up for a toroidal magnetic field during lowering into the cryo-container of an experimental plant at the German Research Center Karlsruhe. The experimental plant serves to develop the technology of magnetic plasma confinement for the nuclear fusion process. The outer dimensions of the oval model coil are 2.55 m × 3.60 m × 0.58 m. During its operation an electric current of 80,000 A is flowing through the coil. The total coil set-up weighs 107 tons and must be cooled down to 4.5 K. The available inner diameter and height of the cryo-container is 4.3 m and 6.6 m, respectively (Research Center Karlsruhe KIT)

Since, in a superconducting coil, direct current can flow without any losses for a practically arbitrarily long time, such coils offer an interesting possibility for the storage of electrical energy, in particular for the handling of short interruptions of the electric power supplied. Therefore, at present the development of superconducting magnetic energy-storage systems is pursued intensively. Nuclear fusion as a long-term option for a source of energy must rely necessarily, for energetic reasons, on superconducting coils to generate the magnetic fields needed for the confinement of high-temperature plasma, in which the nuclear fusion process occurs. Hence, the largest superconducting magnet systems are developed presently for application in nuclear fusion reactors.

Once technical superconductors with their highly improved superconducting material properties became available, then in the 1970s the investigation of the basic principles of superconducting electric power cables, based on the classical superconductors and cooled with liquid helium, had already been started. In this context several pilot projects were carried out worldwide. Today, about 95% of the electric power is transmitted using alternating current high-voltage open-air power lines. Because of their relatively low construction and repair costs, they have distinct advantages. The possible operation of superconducting electric power cables is particularly useful at such locations, where open-air power lines are ruled out as, for example, in regions with very high population density. Because of the discovery of high-temperature superconductivity, which we will discuss in the following Chap. 9, this high-current application of superconductivity has also received a strong impetus.

Chapter 9
The Big Surprise: High-Temperature Superconductivity

Abstract The discovery of high-temperature superconductors in 1986 started worldwide tremendous research activities which quickly resulted in the preparation of superconductors with a critical temperature above 130 K. These cuprate super-conductors are highly anisotropic, with superconductivity residing in the copper-oxide planes. Initially, the granular structure of the ceramic materials needed to be optimized. Much progress was achieved by fabricating epitaxial films. The symmetry of the wave function of the Cooper-pair condensate represents an important issue. Today, the Josephson effect of a single grain boundary is used in SQUIDs. The intrinsic Josephson effect in small multi-layer crystals is explored as a source of terahertz radiation. Major events were the discovery of superconductivity in MgB_2 in 2001 and in iron-pnictides in 2008.

9.1 Cuprate Superconductors

In April 1986 the German Johannes Georg Bednorz together with the Swiss Karl Alexander Müller submitted a paper for publication in the *Zeitschrift für Physik* with the title "Possible High T_C Superconductivity in the Ba-La-Cu-O System". Both worked at the IBM Research Laboratory in Rüschlikon near Zurich. In compounds of barium, lanthanum, copper, and oxygen, with decreasing tempera-ture, they had observed an abrupt drop in the electrical resistance by at least three orders of magnitude, with the drop starting at about 35 K. The two scientists presumed that they were dealing with a new kind of superconductivity. Because the superconductivity appeared to set in at a temperature, which was up to 12 K higher than the highest recorded value of the critical temperature of 23.2 K known at the time (since 12 years) for the compound Nb_3Ge, caution and scepticism was called for. Therefore, the authors arranged with the editor of the *Zeitschrift für Physik* to hold the paper until a clear proof of the superconductivity was provided by an experimental demonstration of the Meissner effect. As we saw in Chap. 8, the Meissner effect represents the characteristic fingerprint of superconductivity. The

© Springer Nature Switzerland AG 2019
R. P. Huebener, *Conductors, Semiconductors,
Superconductors*, Undergraduate Lecture Notes in Physics,
https://doi.org/10.1007/978-3-030-31420-0_9

Meissner effect was then, indeed, confirmed also for the Ba-La-Cu-O system, and Bednorz and Müller released their submitted paper for publication.

Initially, Bednorz and Müller mostly faced scepticism. However, this lasted only a short time. Already by the end of 1986 their results had been confirmed at the University of Tokyo and only a little later at the University of Houston in the American Federal State of Texas. Then in 1987, Paul Ching-Wu Chu, Maw- Kuen Wu, and their co-workers in Houston succeeded in another sensational advance. In a modification of the original oxides, in which the larger lanthanum atom was replaced by the smaller yttrium atom, they observed an enormous increase in the critical temperature up to 92 K. Now the investigations into the "high-temperature superconductors" developed a breathtaking speed worldwide in many groups. The critical temperature of 92 K for the, just-discovered, new material $YBa_2Cu_3O_7$ (abbreviated YBCO) is still distinctly higher than the boiling point of 77 K for liquid nitrogen. Therefore, the relatively expensive liquid helium as a cooling medium can be replaced by the much cheaper liquid nitrogen. In many places the increasingly hectic rush was so great, that temporarily new results were reported in daily newspapers such as, for example, in The New York Times, with an exact indication of the day and the hour they had been achieved. Perhaps the first crucial point was the marathon session at the Spring Conference of the American Physical Society of March 18, 1987 in the Hilton Hotel in New York City, which lasted far beyond midnight and subsequently was accurately called the "Woodstock of Physics".

The discovery of high-temperature superconductivity by Bednorz and Müller resulted in an explosive growth worldwide of research and development in the field of superconductivity. With respect to the international reaction, this discovery can be compared with the discovery of X-rays in 1895 by Wilhelm Conrad Röntgen or with the first observation of nuclear fission in 1938 by Otto Hahn and Fritz Strassmann. It is estimated that, until the beginning of the year 2001 a total of about 100000 scientific papers on high-temperature superconductors had been published, since their discovery (Fig. 9.1).

In particular the lecture given by Bednorz during the ceremony in which the Nobel Prize was awarded to him together with K. A. Müller on December 8, 1987 in Stockholm, gives a vivid and enlightening description of the path leading to their discovery of high-temperature superconductivity. We will quote a few passages from this lecture:

"....We started the search for high-T_C superconductivity in late summer 1983 with the La-Ni-O-System." – Bednorz then talks about various steps, during which the nickel and the lanthanum were replaced by other elements, but without success. Then he continues: "The resistance behavior changed in a way we had already recorded in the previous case, and at that point we started wondering whether the target at which we were aiming really did exist. Would the path we decided to embark upon finally lead into a blind alley?"

"It was in 1985 that the project entered this critical phase, and it probably only survived because the experimental situation, which had generally hampered our efforts, had been improved. The period of sharing another group's equipment for resistivity measurements came to an end. ... Thus the measuring time was transferred from late evening to normal

Fig. 9.1 Critical temperature T_C plotted versus the year of discovery of different superconductors. The steep branch of the curve on the right-hand side shows the different high-temperature superconductors. The entries marked with an asterisk indicate where the critical temperature could be increased further under high pressure. (C. W. Chu)

working hours." – After a brief summary of the further experiments Bednorz continues "…. But again, we observed no indication of superconductivity. The time to study the literature and reflect on the past had arrived."

"It was in late 1985 that the turning point was reached. I became aware of an article by the French scientists C. Michel, L. Er-Rakho, and B. Raveau, who had investigated a Ba-La-Cu oxide with perovskite structure exhibiting metallic conductivity in the temperature range between +300 and −100 °C….. In the Ba-La-Cu oxide with a perovskite-type structure containing Cu in two different valencies, all our concept requirements seemed to be fulfilled. I immediately decided to proceed to the ground-floor laboratory and start preparations for a series of solid solutions. "–Then there occurred a few interruptions of the experiments, and Bednorz continues in his lecture"…. in mid-January 1986, I recalled that when reading about the Ba-La-Cu oxide it had intuitively attracted my attention. I decided

to restart my activities by measuring the new compound. When performing the four-point resistivity measurement, the temperature dependence did not seem to be anything special when compared with the dozens of samples measured earlier. During cooling, however, a metallic-like decrease was first observed, followed by an increase at low temperatures.....
My inner tension, always increasing as the temperature approached the 30 K range, started to be released when a sudden resistivity drop of 50% occurred at 11 K. Was this the first indication of superconductivity?"

"Alex (Müller) and I were really excited, as repeated measurements showed perfect reproducibility and an error could be excluded. Compositions, as well as the thermal treatment, were varied and within two weeks we were able to shift the onset of the resistivity drop to 35 K. This was an incredibly high value compared with the highest T_C in the Nb_3Ge superconductor."

The substances of the discovered new class of the "cuprate superconductors" (Fig. 9.2) are oxides, which crystallographically have perovskite structure. The

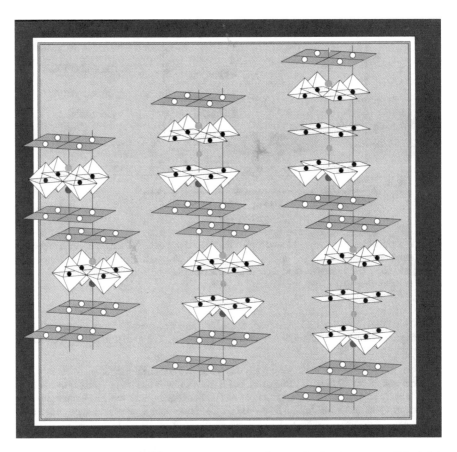

Fig. 9.2 Crystal structure of different cuprate superconductors. At the six corners of the bright octahedrons or at the five corners of the bright pyramids there are oxygen atoms. The centers of the octahedrons, or of the basic square areas of the pyramids, are occupied by copper atoms. (IBM)

Table 9.1 Critical temperatures of different high-temperature superconductors

Compound	T_C (K)
$La_{2-x}Sr_xCuO_4$	38
$YBa_2Cu_3O_{7-x}$	92
$Bi_2Sr_2Ca_2Cu_3O_{10}$	110
$Tl_2Ba_2Ca_2Cu_3O_{10+x}$	125
$HgBa_2Ca_2Cu_3O_{8+x}$	133

most prominent structural element are copper-oxide (CuO_2) planes, in which copper and oxygen atoms are arranged alternately, in this way forming a two-dimensional lattice. The elementary building blocks, from which the cuprate superconductors are assembled periodically in all three spatial directions, (the crystallographic unit cells), contain a different number of copper-oxide planes, depending upon the particular compound. Based upon the different possible chemical composition, one distinguishes between five main families of the cuprate superconductors, the parent compounds of which, together with their critical temperature T_C, are listed in Table 9.1.

The electrical and, in particular, the superconducting properties of these cuprates are determined by the copper-oxide planes and depend sensitively on their doping with electric charge carriers. In their undoped state the cuprates are electrical insulators, in which the elementary magnets of the copper atoms in the CuO_2 planes are alternately oriented opposite to each other. Superconductivity is only observed if the electron concentration in the CuO_2 planes is reduced by means of introducing positive holes into the electronic system ("hole doping"). This hole-doping can be achieved, for example, by the extraction of oxygen. Since, on the other hand, superconductivity only appears within a relatively narrow range of the doping concentration, during material preparation the oxygen concentration must be carefully controlled. The values of the critical temperature given in Table 9.1 correspond to the case of optimum doping with holes. The compound $HgBa_2Ca_2Cu_3O_{8+x}$ with $T_C = 133$ K shows the highest critical temperature observed up to now under normal pressure. Under high pressure the critical temperature of this compound reaches the even higher value of $T_C = 164$ K.

In addition to the cuprates, which become superconducting after hole doping, a few compounds have been found, showing superconductivity only after doping with additional electrons, i.e., with negative charges. However, in this case the doping-concentration range required for superconductivity is narrower, and the critical temperature is much lower, compared with the hole-doped compounds.

The layered crystal structure of cuprate superconductors with the dominating role of the CuO_2 planes (Fig. 9.3), results in an extremely strong dependence of all electrical and thermal transport properties upon the direction within the crystal. For example, in the normal state the electrical resistivity perpendicular to the CuO_2 planes is up to several orders of magnitude larger than it is parallel to these planes. In many respects, the materials show quasi two-dimensional behavior. In the normal state of the cuprates, the temperature dependence of the physical properties

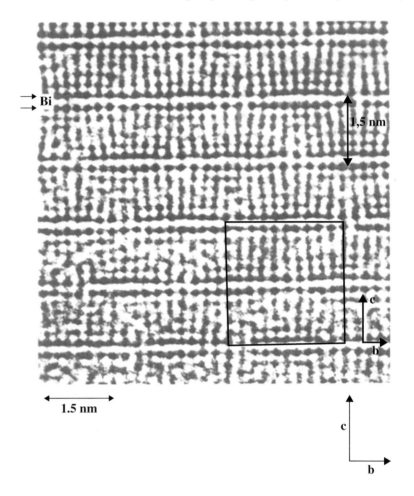

Fig. 9.3 High-resolution electron-microscopic image of the cuprate superconductor $Bi_2Sr_2CaCu_2O_{8+\delta}$. The two arrow-heads at the upper left show two rows of bismuth atoms. At the bottom, the direction of the crystallographic b- and c-axis is indicated. The c-axis is oriented perpendicular to the CuO_2 planes of the cuprate. (O. Eibl)

such as electrical resistance, the Hall effect, as well as the Seebeck and Peltier effects, strongly deviates from the behavior usually observed in metals.

In the high-temperature superconductors, the coherence length ξ, which characterizes the spatial rigidity of the superconducting properties, is much smaller than in the classical superconductors and has a similar magnitude to the dimensions of the crystallographic unit cell. We are dealing with extreme type-II superconductivity. Therefore, these materials are extremely sensitive against atomic defects and grain boundaries, both of which act as pinning centers for magnetic flux quanta (Fig. 9.4). The upper critical magnetic field H_{C2} is up to more than 100–200 times

Fig. 9.4 Granular structure of one of the first prepared thin layers of the cuprate superconductor $Y_1Ba_2Cu_3O_7$. The layer is 30 μm wide and runs horizontally. The arrow-heads on the right-hand side mark the upper and the lower edge of the layer, respectively. Bright regions indicate the locations in which electrical resistance appears within the layer during electric current flow. The dark regions are superconducting. From (a) to (e) the electric current was increased successively from 0.7 mA at (a) up to 8.7 mA at (e). The images show the pronounced spatial inhomogeneity of the layer, having large spatial fluctuations of the local critical electric current density. The images were obtained using the method of low-temperature scanning electron microscopy. The temperature was 53 K

larger than the highest values for the classical superconductors. One of the first questions which had to be answered for the newly discovered materials, concerned the issue of whether the formation of Cooper pairs is fundamental to superconductivity, similar to the classical superconductors. For cuprate superconductors, this question of pair formation had received a clear positive answer early on. Here definite indications for the appearance of the double elementary charge of the Cooper pairs came from the magnitude of the magnetic flux quantum and from the quantitative relation between the electric voltage and the frequency of the Josephson effect.

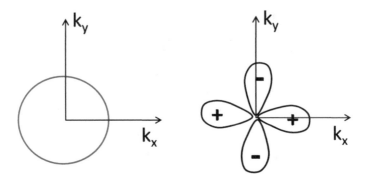

Fig. 9.5 Illustration of the wave function with s-wave symmetry (left) and with $d_{x^2-y^2}$-symmetry (right) in **k**-space ($k_x - k_y$ plane). The latter symmetry dominates in the CuO_2 planes of the cuprate superconductors

9.2   Symmetry of the Wave Function

The spatial symmetry of the wave function, describing the superconducting ground state of the Cooper pairs, represents another important issue with the high-temperature superconductors. In Chap. 5 we noted, that the states of the electrons in the form of waves propagating within the crystal are determined by the wave vector **k**, and saw how these wave vectors build up the three-dimensional **k**-space or momentum space. Since in the cuprates the superconductivity essentially is concentrated in the CuO_2 planes, we can now practically restrict ourselves to two-dimensional momentum space within these planes. Then the question remains: Does the wave function depend on the direction within this momentum space or not? For the classical superconductors, in general there is no such dependence on the direction, and one speaks of the "s-wave symmetry". However, for the cuprate superconductors the situation is different. For the hole-doped high-temperature superconductors a strong directional dependence of the wave function was observed, which is dominated by the atomic d-orbitals.

In order to illustrate this behavior, the wave function is plotted in two-dimensional **k**-space of the CuO_2 planes. Figure 9.5 shows such a polar plot with the four lobes of the d-orbitals, which alternately have a positive and a negative sign and display nodes and antinodes depending on the polar angle. In the case of $d_{x^2-y^2}$-symmetry, the nodes and antinodes are arranged along the indicated crystallographic directions. For comparison, the isotropic wave function with s-wave symmetry, which usually appears in the classical superconductors, is also shown. In order to identify the directions of the nodes and antinodes within the scheme of the CuO_2 planes, in Fig. 9.6 we show the square CuO_2 lattice.

If we start in momentum space with a specific direction of the wave vector and perform, in the direction of the wave vector, a complete rotation around the center of the system of coordinates, then for d-wave symmetry the wave function changes

Fig. 9.6 Scheme of the square CuO$_2$ lattice. The unit cell is marked by the solid line. The lattice constant a is indicated

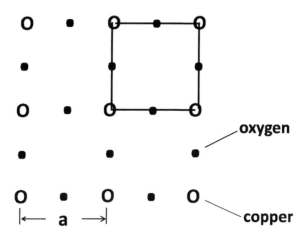

its sign four times until we come back to the starting direction. During this rotation the wave function passes four times through the value zero. These directions with zero value are called nodes, as is common with vibrating strings. At the nodes the energy gap in the superconductor vanishes, and it increases again at both sides of the nodes. For hole-doped high-temperature superconductors, the d-wave symmetry of the wave function of the Cooper pairs, with its sign changes and its nodes, leads to many important consequences in the physical properties of the superconducting state of these materials. In this context, we will discuss below a beautiful experiment for the detection of half-integer magnetic flux quanta. For electron-doped high-temperature superconductors the question of the spatial symmetry of the wave function of the Cooper pairs is not yet completely clarified, since experimental observations do not yet allow an unequivocal conclusion.

Whereas the formation of Cooper pairs can be stated definitely as a fundamental principle also for high-temperature superconductors, the underlying microscopic pairing mechanism still remains unclear for the cuprates and remains a theoretical and experimental challenge.

The layered structure of the cuprate superconductors with the CuO$_2$ planes arranged on top of each other, also affects the magnetic flux lines, which Abrikosov had predicted first for the type-II superconductors, and which in the mixed state intersect the superconductor like a forest of poles along the direction of the magnetic field. We only consider the case where the magnetic field is oriented perpendicular to the CuO$_2$ planes. Since the superconducting property is highly concentrated within these planes, the flux lines are also generated only along a short distance on the planes and are interrupted between the planes. Now the continuous magnetic flux line, according to the theory of Abrikosov, is separated into short disks, which are located at the CuO$_2$ planes, and which are stacked exactly on top of each other. Often these disks are referred to as "pancakes". Because of this separation of the magnetic flux lines into many small individual disks, the magnetic flux-line lattice now displays a large number of new properties, which are absent in the original Abrikosov lattice. For example, individual disks can leave the arrangement where they are stacked exactly on top of each other, and a process like the melting and evaporation of the perfectly-stacked configuration of the disks becomes possible.

In Chap. 8 we discussed that, because of the motion of the magnetic flux lines, an electric voltage is generated in the superconductor, and noted that this leads to electric losses, if the flux-line motion is caused by the Lorentz force produced by an electric current. Because of the decomposition of the flux lines into the individual disks (pancakes), in the high-temperature superconductors, this loss mechanism is particularly important, since the motional freedom of the individual small disks is much stronger than that of the complete and, more or less rigid, Abrikosov flux lines. Therefore, the prevention of the motion of the disks by the introduction of pinning centers into the superconductor represents a highly important task. In this context we recall that the radius of the normal core in the center of each flux line is given by the coherence length ξ, which is much smaller in the cuprates than in the classical superconductors. Hence, the minimum size of the pinning centers only needs to reach about an atomic length scale, in order to be effective. This explains why even only local deviations from stoichiometry, such as, for example, missing oxygen atoms in the CuO_2 planes and grain boundaries on an atomic scale, represent highly effective pinning centers in the high-temperature superconductors.

9.3 Grain Boundaries

Soon after the discovery of the cuprate high-temperature superconductors, a severe problem with these materials became apparent. As ceramics, the materials were prepared initially with a granular structure, where the individual grains were separated from each other by a dense network of grain boundaries. Since, in general, within these grain boundaries the superconductivity is weakened or even interrupted, during electrical current flow a finite electrical resistivity was observed, and hence no pure superconductivity appeared. An early experiment demonstrating the granular structure of the cuprate superconductor $Y_1Ba_2Cu_3O_7$ is shown in Fig. 9.4. This problem defined two obvious goals for additional research and development. On the one hand, methods had to be found for strongly reducing the number of grain boundaries in the material. On the other hand, the physical properties of the grain boundaries themselves had to be investigated exactly.

Regarding the first goal, impressive progress could be achieved relatively quickly. Here above all it was thin-film technology which allowed the preparation of thin, single-crystalline layers of the high-temperature superconductors deposited on suitable substrates. These "epitaxial layers" contain hardly any grain boundaries and remain clearly superconducting up to critical electric current densities of more than one million A/cm^2 at the temperature of 77 K, the boiling point of liquid nitrogen.

The pursuit of the second goal, namely the clarification of the physical properties of the grain boundaries, resulted then in an unexpected but highly interesting development. Here it was mainly scientists at the Thomas J. Watson Research Center of IBM in Yorktown Heights in the American Federal State of New York, which dominated this development. In order to study the physical behavior of a single grain boundary, in his IBM laboratory Chang C. Tsuei selected a thin layer of the high-temperature superconductor $Y_1Ba_2Cu_3O_7$ showing relatively large single-crystalline areas separated from each other by very long individual grain boundaries. Now only a narrow conducting bridge had to be fabricated out of the YBCO layer in such a way, that the bridge was running nearly perpendicular across the grain boundary. This allowed electrical measurements to be performed on a single grain boundary. During the time when Tsuei prepared his experiments using a spontaneously generated single grain boundary, his IBM colleague Praveen Chaudhari proposed the idea of generating the grain boundary in a controlled way by means of a specially prepared substrate for depositing the layer of the cuprate superconductor. During the further developments, also the postdoc Jochen Mannhart was heavily involved. With the method used by Tsuei for achieving the epitaxial growth of the single-crystalline layer of the high-temperature superconductor, the crystallographic orientation of the single-crystalline substrate is reproduced exactly by the superconducting layer deposited on top. If for the substrate one uses an artificially prepared so-called bicrystal, in which two single-crystalline parts with different crystallographic orientation are separated from each other by an atomically sharp grain boundary, the grain boundary of the substrate is transferred exactly to the superconducting layer on top. This bicrystal is fabricated by cutting up a single-crystal into proper pieces and by fusing two pieces, with a different crystal orientation, together again (Fig. 9.7). The bicrystal technique turned out to be highly successful and subsequently permitted many experiments with well-defined grain boundaries in superconducting thin layers. In particular, the Josephson effect at a single grain boundary was observed. This technique then developed into an important method for the preparation of Josephson junctions in thin layers of high-temperature superconductors. Furthermore, this approach also served extremely well for the fabrication of SQUIDs based on high-temperature superconductors.

The different signs of the $d_{x^2-y^2}$-wave function appearing at different polar angles, respectively, can lead to important consequences, if two crystals with different orientation are connected with each other by means of a well-defined grain boundary (bicrystal technique, Fig. 9.7). The case, at which a positive lobe of the wave function, at the opposite side of the junction runs into a negative lobe, is referred to as π-junction. A closed ring, containing such a π-junction, violates the uniqueness of the wave function (frustration), since, during a complete revolution, a sign change of the wave function remains. In this case, the frustration is removed by the spontaneous generation of a half-integer magnetic flux quantum.

Eventually, an extension of this principle was utilized by Chang Tsuei for proving experimentally the d-wave symmetry of the Cooper pair wave function in

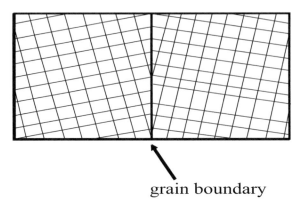

grain boundary

Fig. 9.7 Bicrystal technique for the controlled preparation of a single grain boundary within a superconducting cuprate layer. As the substrate one uses an artificially prepared bicrystal, in which two differently oriented single-crystalline parts of the crystal are separated from each other by an atomically sharp grain boundary. The grain boundary within the substrate is then exactly transferred to the superconducting layer on top. On both sides of the grain boundary there now exist single-crystalline superconducting layers with different crystallographic orientation

hole-doped high-temperature superconductors (Fig. 9.8). We discussed above that, for d-wave symmetry, the wave function changes its sign four times during a complete rotation of the direction of the wave vector. Therefore, each time after only half a full rotation, the same sign of the wave function appears again. By joining together three angular sections like the pieces of a round cake in each of which the crystallographic orientation of the superconducting layer is different, it can be achieved that at one of the three grain boundaries a sign change of the wave function between the two sides takes place. If this is the case, after a complete rotation of the direction around the common meeting point of the three angular sections, the wave function no longer reproduces itself, but instead a sign change remains due to the generation of a π-junction. As a necessary consequence, at the common meeting point of the three grain boundaries, an exactly half-integer magnetic flux quantum is spontaneously generated. Chang C. Tsuei, and collaborators have succeeded in detecting this half-integer magnetic flux quantum by means of a SQUID-scanning microscope. The fabrication of the three angular sections of the superconducting layer, each with a different crystal orientation and separated from each other by sharp grain boundaries, could be achieved using a "tricrystal" as a substrate, where this tricrystal was composed of three correspondingly oriented angular sections. In the case of the tricrystal, the cutting and fusing process for producing the bicrystals had to be extended accordingly. This tricrystal experiment and the detection of the spontaneously generated half-integer magnetic flux quantum at the common meeting point of the three grain boundaries, represents one of the most spectacular demonstrations proving the d-wave symmetry of the pair wave function in hole-doped high-temperature superconductors.

Fig. 9.8 Tricrystal experiment of Tsuei and coworkers for proving the d-wave symmetry of the quantum mechanical wave function of Cooper pairs in the cuprate superconductor $Y_1Ba_2Cu_3O_7$. The substrate is an artificially prepared tricrystal, in which three differently oriented single-crystalline parts of the crystal are separated from each other by atomically sharp grain boundaries. This crystalline structure, including its grain boundaries, is exactly transferred to the superconducting layer prepared on top. The grain boundaries are marked by straight white lines. Within the three crystal parts, separated from each other by the grain boundaries, the differently oriented d-wave symmetry pattern of the Cooper pair wave function is indicated by the diagrams with the four lobes. At different locations, a total of four superconducting rings are fabricated from the YBaCuO layer, whereas the remaining part of the layer is removed. The orientations of the three crystal parts are chosen such that, in the presence of d-wave symmetry of the wave function, within the ring around the common meeting point of the three crystal parts, an exactly half-integer magnetic flux quantum is spontaneously generated, whereas nothing happens at the three other rings. The image was obtained by means of a SQUID scanning microscope, and it clearly shows the half-integer magnetic flux quantum for the ring in the middle around the common meeting point of the three crystal parts. The other rings are only weakly visible. (C. C. Tsuei)

9.4 Intrinsic Josephson Junction

The crystal structure of the cuprate superconductors suggests a Josephson contact, in which the superconducting copper-oxide planes are separated from each other by intermediate layers, which are only weakly electrically conducting. In a crystal of these materials there are up to many thousand Josephson junctions of this type stapled upon each other. In 1992, Reinhold Kleiner and Paul Müller, at the time working at the Walther-Meissner Institute for Low Temperature Research of the Bavarian Academy of Science in Garching near Munich, have succeeded in discovering the "intrinsic Josephson effect" in small single crystals of the cuprate superconductor $Bi_2Sr_2CaCu_2O_8$ (BSCCO).

During the initial experiments a small superconducting BSCCO-crystal was clamped between two contact pins such that an electric current flow perpendicular to the copper-oxide planes occurred. Above a critical value of the current, an electric voltage appeared along the crystal which, according to the second Josephson equation (8.19), was accompanied by high-frequency Josephson

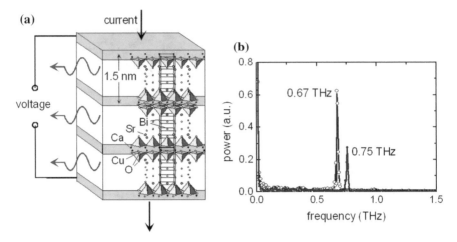

Fig. 9.9 Intrinsic Josephson contact as a source of microwaves. Left: Scheme of a staple of three Josephson contacts of a superconducting $Bi_2Sr_2CaCu_2O_8$ (BSCCO) crystal. The copper-oxide planes (shown in dark color) run along the base of the CuO-pyramids. Right: Emitted microwave spectrum of a BSCCO-crystal. (R. Kleiner)

alternating current. In particular, the emitted microwaves could be detected. Since the stapled Josephson contacts oscillate synchronously, the emitted power of the electromagnetic radiation increases quadratically with the number of contacts in the staple. Presently, for the study and application of the intrinsic Josephson effect, small "mesas" made of BSCCO are prepared on a suitable substrate and are covered by metal layers for electrical contacting (Fig. 9.9). Currently, these developments are highly active aiming at a radiation source of microwaves in the frequency range of about $0.5 - 2$ terahertz (THz $= 10^{12}$ s^{-1}), since up to now this frequency range is hardly accessible (terahertz gap). Presently (2019) the power output of a single mesa is up to 0.2 mW, whereas for a synchronized array of several mesas it amounts up to 0.6 mW. This limit is mainly due to electric heating effects within the device.

9.5   More New Superconductors: MgB$_2$, Iron Pnictides

The surprises caused by the discovery of new superconducting materials with relatively high values of the critical temperature were not yet over with the appearance of the cuprate superconductors. In March 2001 the group of Jun Akimitsu at the Aoyama-Gakuin University in Tokyo reported in the journal "Nature", that the compound MgB$_2$, consisting only of two elements, is super-conducting with the critical temperature $T_C = 39$ K. Already by January 2001 Akimitsu had announced his discovery at a conference in Japan, and immediately many groups started research activities to find out the physical properties of this new superconductor. Similar to the cuprates, the crystal structure of the magnesium-diboride (MgB_2) shows a layered configuration. Planes of hexagonally arranged magnesium atoms and planes of boron atoms ordered in a honeycomb pattern, like graphite, are placed alternately on top of each other. As expected it was again found, that the superconductivity is based on the formation of Cooper pairs. Similar to the classical superconductors, the wave function of the pairs does not show an appreciable dependence on the direction, apparently displaying s-wave symmetry.

Early in 2008, the group of Hideo Hosono in Japan reported another surprise: In the compound LaOFeAs of lanthanum (La), oxygen (O), iron (Fe), and arsenic (As), they had discovered superconductivity, following its doping with fluorine (F). In the case of the composition $LaO_{1-x}F_xFeAs$ at x = 0.07 the critical temperature T_C = 26 K was found. This new class of superconductors containing iron and arsenic belongs to the group of the iron pnictides.

In the family $ReO_{1-x}F_xFeAs$ additional superconductors were found rather quickly, after lanthanum was replaced by other elements of the rare earths (RE) such as praseodymium (Pr), neodymium (Nd), or samarium (Sm). In this way, values of the critical temperature up to the record value T_C = 56 K of $Sr_{0.5}Sm_{0.5}FeAsF$ could be achieved. In spite of their pronounced layered structure, in their electronic transport properties the iron-pnictides do not show appreciable anisotropy (like the cuprates). The important structural elements are planar layers of iron atoms surrounded by tetrahedrically coordinated As or Se anions playing the role of the CuO planes in the cuprates (Fig. 9.10). The layers are arranged in a stacked sequence separated by alkali, alkaline-earth, or rare-earth and oxygen blocking layers, in which oxygen is replaced partly by fluorine acting as dopant.

Similar to the case of the cuprates 22 years earlier, worldwide the research dealing with the iron-pnictides grew explosively. Similar to the cuprates, in their undoped state the iron- pnictides are ordered magnetically, but they are electrically conducting contrary to the cuprates. They are an anti-ferromagnetic semi-metal. It appears, that in this case superconductivity and magnetism are connected with each other. An overview of the different iron-pnictides discovered up to 2015 is shown in Fig. 9.11. The different families are distinguished by the notation 11, 111, 122, 1111, etc., respectively. By 2010 more than 50 iron-pnictide compounds had been found. The pairing mechanism remains unclear. However, a large amount of evidence points to magnetic spin fluctuations.

Fig. 9.10 Layer of FeAs (or similarly of FeSe) arranged between layers of lanthanum-oxide perhaps doped with fluorine

Fe
As

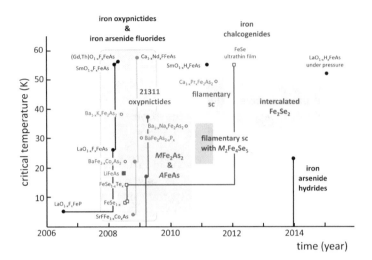

Fig. 9.11 Critical temperature T_C plotted versus the time of discovery of the different iron-pnictide superconductors (Silvia Haindl)

9.6   Technical Applications

Because of the relatively high values of their critical temperature, compared with the classical superconductors (with the possibility of utilizing superconductivity after cooling down to only 77 K with liquid nitrogen) high-temperature superconductors quickly became very useful for technical applications. Here the applications in the field of electronics and microelectronics, as well as the applications at high electric currents and in power electronics appear equally interesting. The following examples will serve as an illustration. Today, the principle of the bicrystal substrates for the fabrication of Josephson grain-boundary junctions and SQUIDs from thin layers of high-temperature superconductors has already found wide applications in electronic measuring instruments. High-frequency filters fabricated from thin layers of high-temperature superconductors appear very promising. Here it is in particular the increased sharpness of the high-frequency channels, achieved with the superconducting layers, which allows many more channels in the available frequency bands to be accommodated than in the past. For example, more than a few hundred base stations for mobile telephone communication, based on this technology, are already operating in the USA. However, following a rapid growth of this superconducting technology, advances in the newer generations of mobile telephones eliminated the need for such superconducting filters, and their acceptance decreased.

Regarding the applications at high electric currents, the development of magnetic coils fabricated from high-temperature superconductors is being intensively investigated. Last but not least, superconducting systems for the limitation of electric fault currents in energy technology are in a promising stage of development.

Such systems are meant to quickly interrupt the electric current under overload conditions, if damages to the electric power lines are expected due to overload.

Presently (2013) an interesting new development based on wind energy emerges in the case of generators made of high-temperature superconductors for electric currents. Their intended operation would reduce the weight of the generators located at the upper end of the mast to about half the value of the current technology, or at the same weight it would double the power.

In Japan, the development of a superconducting dc cable cooled with liquid nitrogen and projected for 5 GW power (Ishikari Project) is taken up.

Chapter 10
Magnetism: Order Among the Elementary Magnets

Abstract Electrons represent elementary magnets because of their spin magnetic moment, leading to paramagnetism, and their orbital magnetic moment, causing diamagnetism. In an external magnetic field the magnetization of a paramagnet is described classically by the Langevin function, to be replaced by the Brillouin function in the case of quantization effects. In a ferromagnet the elementary magnets are spontaneously oriented along a distinct direction. The first quantum mechanical explanation was proposed by Heisenberg, based on the exchange interaction between the electrons of two atoms. The perfect order of the elementary magnets can be perturbed due to the thermal excitation of spin waves, which also contribute to the specific heat and affect the electronic transport properties. In addition to ferromagnetic order, other forms of the spin magnetic order are possible, such as antiferromagnetism. The recent advances in the fabrication of well-controlled multilayer structures lead to important technical applications of giant magneto-resistance and to the birth of the new field of spintronics.

In crystals, the role of the electrons as elementary magnets leads to important consequences, which we will discuss in this chapter. This quality of the electrons to act as elementary magnets, originates for two reasons: the angular momentum or spin possessed by each electron, and the orbital momentum resulting from the orbital motion of each electron. The first cause leads to the spin magnetic moment, and the second cause to the orbital magnetic moment. The half-integer spin and the associated magnetic moment as fundamental properties of the electrons were found theoretically by the Englishman Paul Adrien Maurice Dirac in the year 1928, when he applied the physics of the Special Theory of Relativity to the quantum mechanical wave equation of the electron. Because of the quantization of the direction of the angular momentum, in a given direction only the parallel or the antiparallel orientation of the spins are allowed. With his concept of spin, Dirac had found the explanation for many experimental observations, which had remained unexplained till then, and which in the words of Wolfgang Pauli at that time had

© Springer Nature Switzerland AG 2019
R. P. Huebener, *Conductors, Semiconductors, Superconductors*, Undergraduate Lecture Notes in Physics,
https://doi.org/10.1007/978-3-030-31420-0_10

indicated in the energy spectra (here in the English translation) *"a peculiar ambi-guity of the quantum theoretical qualities of the light-emitting electron, which cannot be understood classically."*

10.1   Diamagnetism

The diamagnetism is due to the orbital magnetic moment, and is a magnetic property of all substances. However, it can be overlaid by other magnetic phe-nomena. Diamagnetism appears in its pure form, if the spin magnetic moments of all atomic electrons exactly compensate each other, such that only the orbital magnetic moments remain. This complete compensation arises in the case of atoms with closed inner electronic shells and of atoms with an even number of electrons. In an external magnetic field, in diamagnetic materials, circulating currents are induced within the atoms, which generate a magnetic field oriented oppositely to the external magnetic field. Here the "Lenz rule" is obeyed. This rule requires, that the induced electric currents always weaken the magnetic field which is acting as their external cause. Therefore, diamagnetic susceptibility is negative. The orbital part of the magnetism of the electrons in the conduction band of a metal yields a diamagnetic contribution, which Lew Dawidowitsch Landau calculated for the first time in the year 1930, using exact quantum mechanical theory. We have discussed this theory of Landau in Chap. 7. We have already seen an example of perfect diamagnetism in the context of the Meissner effect of superconductivity.

The diamagnetism is a result of the change of the orbital motion of the electrons due to an external magnetic field and concerns only the magnetic moments caused by the magnetic flux density **B**.

The magnetic properties of a material are quantified in terms of its magnetization **M**, which is defined as the magnetic moment per volume. In the case of **M** we have the relation

$$\mathbf{M} = \chi\mathbf{B}, \tag{10.1}$$

where χ denotes the magnetic susceptibility. In the case of diamagnetism we have $\chi < 0$. In a magnetic field, the electron orbits experience a precession motion with the Larmor frequency

$$\omega_L = eB/2m. \tag{10.2}$$

Here m is the electron mass. The Larmor precession corresponds to a circular current proportional to $(-e)\,\omega_L$. This leads to the magnetic moment μ in form of the product $\mu =$ (circular current × ring area). As shown by Paul Langevin, one finds for the (negative) susceptibility

$$-\chi \sim <r^2> , \qquad (10.3)$$

where $<r^2>$ denotes the average quadratic radial extension of the electron orbit within the atom. The diamagnetic contribution of the ions to the susceptibility in dielectric solids agrees well with the result (10.3).

10.2   Paramagnetism

If the electron shells are not closed, or for an odd number of electrons per atom, there remain in the crystal spin magnetic moments, which are not completely compensated. In this case we are dealing with paramagnetism. Initially, in para-magnetic materials, the spin magnetic moments are completely disordered. However, as soon as an external magnetic field exists, the spin moments turn into the direction of this magnetic field. Here the electrons behave similarly to a com-pass needle, which turns into the direction of the earth's magnetic field. However, the complete redirection of the spin magnetic moments is prevented because of the thermal motion of the elementary magnets. This thermal motion of the elementary magnets increases with increasing temperature. Hence, the degree of redirection of the spin magnetic moments decreases correspondingly. This results in the famous Curie law, indicating that the magnetic susceptibility, which is positive in this case, is inversely proportional to the temperature. The Frenchman Pierre Curie was married to Marie Curie, who had discovered the element radium. He published his law in the year 1895. This was the first time that a law in the field of the magnetism of materials had been formulated. Curie's publication with more than one hundred pages summarized the results of an extensive research program, in which many substances had been investigated over a large range of magnetic field and tem-perature. Curie's name stands at the beginning of a series of French scientists, who had made France, at an early stage, an important center of research in the field of magnetism. Among others, this series of French scientists prominent in the field of magnetism includes Paul Langevin, Pierre Weiss, Léon Brillouin, and Louis Eugene Felix Néel.

Paul Langevin, a young co-worker of Pierre Curie, analyzed theoretically paramagnetic behavior. We will briefly sketch his ideas. Starting from the potential energy

$$U = -\boldsymbol{\mu} \cdot \mathbf{B} = -\mu B \cos \theta \qquad (10.4)$$

of a magnetic moment $\boldsymbol{\mu}$ in the magnetic field $\mathbf{B}$, Langevin calculated the average value $<\cos \theta>$ from the classical Boltzmann distribution in the following way. (θ is the angle between the vectors $\boldsymbol{\mu}$ and $\mathbf{B}$). According to Boltzmann, the probability of the direction of the magnetic moment $\boldsymbol{\mu}$ with the corresponding energy U is pro-portional to $\exp(-U/k_B T)$. Therefore, (with the solid angle $d\Omega$) we have

$$<\cos \theta> \; = \frac{\int \exp\left(-\frac{U}{k_B T}\right) \cos \theta \, d\Omega}{\int \exp\left(-\frac{U}{k_B T}\right) d\Omega} \tag{10.5}$$

$$= \frac{\int_0^\pi \exp(\mu B \cos \theta / k_B T) 2\pi \sin \theta \cos \theta \, d\theta}{\int_0^\pi \exp(\mu B \cos \theta / k_B T) 2\pi \sin \theta \, d\theta} \tag{10.6}$$

and after a few steps

$$<\cos \theta> \; = \coth x - \frac{1}{x} \equiv L(x); \quad x = \mu B / k_B T. \tag{10.7}$$

$L(x)$ is referred to as Langevin function. For the average magnetization $<M>$ one finds

$$<M> \; = N\mu <\cos \theta> \; = N\mu L(x), \tag{10.8}$$

where N denotes the number of the magnetic moments μ per volume. The function $L(x)$ is shown in Fig. 10.1.

In the case $x \ll 1$ (high temperatures) we have $L(x) \approx x/3$, and we obtain

$$<M> \; = N\mu^2 B / 3 k_B T = \frac{C}{T} B. \tag{10.9}$$

This is Curie's law. $C = N\,\mu^2/3\,k_B$ is the Curie-constant. On the other hand, in the case $x \gg 1$ (low temperatures) we obtain saturation of the magnetization.

From (10.7)–(10.9) we see, that the dependence of the magnetization on the temperature and on the magnetic field is determined only by the ratio of the magnetic field and the temperature. In the limit of small magnetic fields and high temperatures, Langevin obtained again the Curie law, whereas in the opposite limit he found that the magnetization approaches a constant value.

Langevin had obtained his result (10.8) still within the context of classical physics. If the quantization of the spin direction, required by quantum mechanics, is

Fig. 10.1 Langevin function $L(x)$ according to the definition in (10.7)

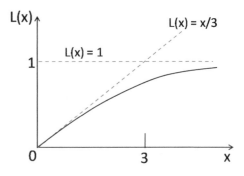

taken into account, the results are qualitatively similar to the classical case. We will briefly indicate this. As shown by quantum theory, the magnetic moment of an isolated atom is

$$\boldsymbol{\mu} = -g\,\mu_B \mathbf{J}/\hbar \tag{10.10}$$

Here

$$\mu_B = e\,\hbar/2\,m \tag{10.11}$$

is Bohr's magneton. $\mathbf{J}$ is the total angular momentum, the sum of the vectors of the angular momentum of the electron orbits ($\mathbf{L}$) and of the spins ($\mathbf{S}$):

$$\mathbf{J} = \mathbf{L} + \mathbf{S}. \tag{10.12}$$

In (10.10) g denotes the "Landé-g-factor":

$$g = 1 + \frac{J(J+1) + S(S+1) - L(L+1)}{2J\,(J+1)}. \tag{10.13}$$

In a magnetic field, the quantized energy levels of an elementary magnetic moment are

$$U = m_J g\,\mu_B B; \quad m_J = J, J-1, J-2, \ldots, -J. \tag{10.14}$$

In the case of a single spin and $L = 0$ we obtain $m_J = \pm\,\tfrac{1}{2}$ and $g = 2$:

$$U = \pm\mu_B B. \tag{10.15}$$

In thermal equilibrium, the magnetization of this two-level-system is

$$<M> \;= N\,\mu_B \tanh x; \quad x = \mu_B B/k_B T. \tag{10.16}$$

In the limit $x \ll 1$ (high temperatures) we find

$$<M> \;= N\,\mu_B^2 B/k_B T, \tag{10.17}$$

similar to Curie's law.

In the general case with $2J + 1$ energy levels according to (10.14), in thermal equilibrium the magnetization is

$$<M> \;= N\,\mu_B B_J(x); \quad x = \mu_B B/k_B T. \tag{10.18}$$

The function $B_J(x)$ is the Brillouin function (which here we will not discuss any further). In the limit $x \ll 1$ (high temperatures) we obtain again the form of Curie's law $<M> \sim B/T$.

In the year 1905, Langevin had already predicted the magneto-caloric effect in paramagnetic substances. This effect is fundamental to the method of adiabatic demagnetization we discussed in Chap. 1 as a technique for cooling to low temperatures. Here the heat energy exchanged during the redirection of the magnetic moments of the electrons parallel to the external magnetic field, plays a central role.

In the conduction band of a metal, the spin magnetic moment of the electrons also generates a contribution to the paramagnetism. Classically, we would again expect a behavior according to the Curie law, with the result that the paramagnetic susceptibility is inversely proportional to the temperature. However, because of the Pauli exclusion principle, the electrons in the conduction band must obey the rules of quantum statistics. As we have discussed before in Chap. 5, therefore, only the fraction of electrons in the conduction band, given by the reduction factor $k_B T/\varepsilon_F$, can contribute to the paramagnetism. By multiplying the Curie law with the factor $k_B T/\varepsilon_F$, we obtain

$$<M> \; = N \mu_B^2 B/\varepsilon_F. \tag{10.19}$$

We see that the temperature in the expression of the paramagnetic susceptibility is cancelled, and that the latter quantity is independent of the temperature, in good agreement with experimental observations. This result is similar to the case of the specific heat of the conduction electrons, which we discussed in Chap. 5. Wolfgang Pauli was the first to propose this theory of the spin paramagnetism of electrons in metals.

In a detailed analysis, one finds for the electrons in the conduction band of a metal, that the negative contribution to the magnetic susceptibility which is due to diamagnetism according to Landau, is three times smaller than the positive contribution which is due to paramagnetism according to Pauli. Hence, in the final result, paramagnetism prevails for the electrons of the conduction band in metals.

10.3 Ferromagnetism

In addition to the phenomena of diamagnetism and paramagnetism, both of which are induced by an external magnetic field, there exists still another form of magnetism, in which the elementary magnets in the crystal orient themselves spontaneously along a distinct direction. In this case, the command by an external magnetic field to order is no longer needed. Now we are dealing with ferromagnetism and its various modifications. For example, one of these modifications is antiferromagnetism. The ferromagnetism confronts us with the following central question: Why is it that the magnetic moments of the electrons of neighboring atoms within the crystal orient themselves spontaneously, i.e., without an external magnetic field, exactly along the same direction and assume a perfectly ordered state just on their own?

The complete theory of ferromagnetism had to wait for the development of quantum mechanics. The crucial answer, which subsequently also became the guiding principle for all further developments, was given by Werner Heisenberg in the year 1928. Heisenberg's answer was preceded by an intensive correspondence with Wolfgang Pauli for almost two years. The basic idea of Heisenberg again originated from the exact identity of the electrons as elementary particles and that the resulting symmetry requirement should be satisfied by the quantum mechanical wave function during the exchange of two electrons. Taking the crystal as an extended molecule, at the time Heisenberg could utilize the concept of the exchange energy, which had just been developed, for, say the two electrons of the helium atom or of the hydrogen molecule (H_2). He introduced the concept of the exchange interaction between two atoms 1 and 2 with the spins S_1 and S_2 and obtained the exchange energy (Heisenberg model)

$$U = -2\,J S_1 \cdot S_2 \tag{10.20}$$

with the exchange integral

$$J = \iint d\,\mathbf{r}_1 d\,\mathbf{r}_2 \psi_a^*(\mathbf{r}_1)\psi_b^*(\mathbf{r}_2)V(\mathbf{r}_1 - \mathbf{r}_2)\psi_a(\mathbf{r}_2)\psi_b(\mathbf{r}_1). \tag{10.21}$$

In the case of ferromagnetism, we have $J > 0$, and the parallel spin orientation is energetically favored. The exact theoretical calculation of the exchange integral (10.21) requires a detailed discussion (which we do not undertake).

For the first time, in this context, Heisenberg treated the interaction between the electrons in a crystal using quantum mechanics. It became clear that the parallel orientation of the spin magnetic moments of the electrons from neighboring atoms in the crystal, leading to the ferromagnetism, depends on the form of the electron wave function and on the number of nearest neighbors in the crystal lattice. At that time the mathematical formalism needed for answering many of the questions, was still only in its first stage and had to be developed along with the quantum theory of ferromagnetism. For this, Heisenberg had given the initial momentum.

A phenomenological understanding of ferromagnetism had already been reached earlier. In the year 1907 the Frenchman Pierre Weiss had proposed the hypothesis of the molecular field or the exchange field, which quickly became highly successful. Without explaining its microscopic origin, he had postulated an average effective magnetic field within the crystal, producing the exact order among the elementary magnets. In this way, the idea of the quantum mechanical exchange energy, appearing only about 20 years later, was substituted by the concept of the effective "Weiss field". It is exactly this effective magnetic field, which causes the spatial order of the magnetic moments of the individual atoms or molecules. Based on the material data, for some substances one can derive values of the Weiss field which are distinctly higher than 10^3 T (about ten million times higher than the earth's magnetic field). Also in this way the large magnitude of the quantum mechanical exchange energy of the magnetism, conceived later, can be illustrated.

During his PhD thesis Weiss concentrated on magnetism, and later always remained close to this subject. As a child, together with his family, he had to leave his home land of Alsace, after the province had been occupied by Prussia in 1870. In the year 1902 he had accepted an offer from the ETH in Zurich. Only in 1919 after the First World War could he return to Strassburg in his home country, when he became the director of the Physics Institute of the University.

In the case of the Weiss field B_W we assume $B_W = \lambda M$, where λ is a constant, which depends on temperature. In a cooperative way, each individual magnetic moment feels the average magnetization of all the others (molecular-field approximation). However, the spontaneous alignment of the elementary magnets along the same direction in a ferromagnet cannot be maintained up to arbitrarily high temperatures. Instead, it vanishes abruptly at the "Curie temperature" T_{CU}. Above the temperature T_{CU} we only have paramagnetism.

In the paramagnetic state (above T_{CU}), in the presence of an external magnetic field B_a, we have

$$M = \chi_p(B_a + B_W) = \chi_p(B_a + \lambda M) \qquad (10.22)$$

with the paramagnetic susceptibility $\chi_p = C/T$ from (10.9). From (10.22) we obtain

$$M(1 - \frac{C}{T}\lambda) = \frac{C}{T}B_a \qquad (10.23)$$

and

$$\chi = M/B_a = C/(T - T_{CU}); \quad T_{CU} \equiv C\lambda; \quad (T \to T_{CU} \text{ from above}) \qquad (10.24)$$

A more accurate treatment yields

$$\chi = C/(T - T_{CU})^{1.33}; \quad (T \to T_{CU} \text{ from above}) \qquad (10.25)$$

with the "critical exponent" 1.33, in agreement with experiment. For the temperature dependence of the magnetic susceptibility, instead of the Curie law, the Curie-Weiss law is valid: the magnetic susceptibility is inversely proportional to the temperature distance $T - T_{CU}$ from the Curie temperature.

Below the Curie temperature, the spontaneous magnetization is found again using the molecular-field approximation $B_W = \lambda M$. In the case of a two-level system ($S = \frac{1}{2}$), by insertion into (10.16), one obtains

$$M = N \mu_B \tanh(\mu_B \lambda M/k_B T). \qquad (10.26)$$

After introducing the quantities $m \equiv M/(N \mu_B)$ and $t \equiv T/T_{CU} = k_B T/(N \mu_B^2 \lambda)$ we obtain

Fig. 10.2 Schematical presentation of the graphical solution of the transcendental (10.27). At the critical point $t = 1$, the intersection point is located at $m = 0$. In the case $t \to 0$, the intersection point approaches $m = 1$

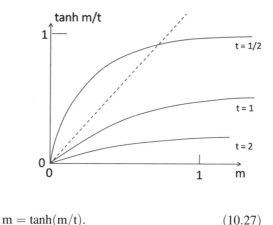

$$m = \tanh(m/t). \qquad (10.27)$$

(Here we have used the notation $T_{CU} = C\,\lambda$, with $C = N\,\mu_B^2/k_B$ from (10.17).) The transcendental (10.27) can be solved graphically, as it is indicated schematically in Fig. 10.2.

The graphical solution yields the temperature dependence of the magnetization shown in Fig. 10.3. This is the behavior in the case of a phase transition of second order, where the magnetization represents the order parameter. From Fig. 10.3 we see, that below T_{CU} the spontaneous magnetization steeply increases with decreasing temperature, and at low temperatures it approaches a constant saturation value. Iron (Fe), cobalt (Co), and nickel (Ni) are well known ferromagnetic elements. The values of the Curie temperature are for iron, 1043 K; for cobalt, 1390 K; and for nickel, 630 K.

Eventually, a difficulty in the understanding of the ferromagnetic order of elementary magnets became more and more apparent: at low temperatures the magnetization turned out to be much smaller than expected if all elementary magnets in the whole crystal were oriented exactly along the same direction. Again, the solution of this problem was provided mostly by Pierre Weiss. Due to energetic reasons the crystal is divided into many individual regions, in each of which all elementary magnets are still well ordered and point exactly in the same direction.

Fig. 10.3 Magnetization of a ferromagnet, normalized by its saturation value, plotted versus the reduced temperature T/T_{CU}. The magnetization vanishes abruptly at the Curie temperature T_{CU}

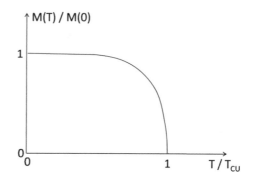

Fig. 10.4 Magnetic domains
according to Weiss in
non-magnetized iron,
schematically

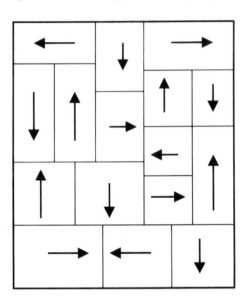

However, between the different regions, the magnetization of each shows a different orientation, such that in their total sum they largely cancel each other. For the individual regions Weiss introduced the notation "magnetic domains" (Fig. 10.4). In beautiful experiments during the year 1931 the American Francis Bitter observed the boundary regions between the domains by sprinkling a fine magnetic powder onto the surface of a magnetized sample. Since the magnetic powder is attracted to these boundary regions, the domain structure is marked in this way. Today this method is referred to as the Bitter decoration technique. In a different way, about 10 years earlier, Heinrich Barkhausen had obtained impressive indications of the existence of the magnetic domains in ferromagnetic substances. While increasing the magnetic field he observed that the magnetization increased discontinuously, showing distinct small jumps when one domain after another reorients its magnetization in the external magnetic field. He could detect these "Barkhausen jumps" by means of the induced and amplified electric currents in a coil wound around the sample.

Immediately after completion of his Ph.D. thesis, Felix Bloch theoretically analyzed the physical property of the boundary wall separating two magnetic domains with different directions of their magnetization. In this context he had to develop a model describing the rotation of the direction of magnetization within the domain wall from the direction in one domain to the direction in the other. Based on Heisenberg's concept of the exchange energy of two neighboring spin magnetic moments, Bloch calculated the energy needed for rotating the two spin magnetic moments slightly away from their exactly parallel orientation. This rotation is then repeated stepwise from one magnetic spin pair to the next, such that, after a distinct number of steps, a complete rotation of the magnetization from the original direction to the direction in the neighboring domain is accomplished. The region within the crystal, in which this complete rotation takes place, is referred to as the Bloch wall. The Bloch wall is associated with a distinct wall energy. For example, in iron, the thickness of the Bloch wall amounts to about 300 atomic distances in the crystal lattice.

10.4 Spin Waves

In the state with the lowest energy, the "ground state", of a ferromagnet, all spins are oriented exactly parallel to each other. However, the ground state is adopted only in the limit of vanishing temperature. At finite temperatures, deviations from the exactly parallel spin orientation appear in the form of thermally excited "spin waves" (Fig. 10.5). Spin waves are oscillations of the spin orientation, which possess a quantized excitation energy $\hbar\omega$. The role of the spin waves is similar to that of the phonons, which are the quantized lattice vibrations and cause the deviations from the perfectly periodical spatial arrangement of the atomic or molecular building blocks of the crystal. We have covered phonons in Chap. 3. Spin waves are quantized energetic excitations in a magnetic single crystal. The energy quanta of the spin waves are called magnons. In their form of energetic excitations, magnons are indistinguishable elementary particles, which, similar to the phonons, are ruled by Bose-Einstein statistics. Intuitively, magnons represent more or less pronounced deviations of the spin magnetic moments from a fixed single preferential direction. These deviations propagate like a wave through the crystal. This concept of the spin waves in the theory of ferromagnetism also originated from Felix Bloch and can be found in his Habilitation Thesis which he published in 1932. Following his dissertation dealing with the quantum mechanics of mobile electrons in the crystal lattice, Bloch had turned to the theory of ferromagnetism, after Werner Heisenberg, his professor, had formulated the fundamental principles of this theory. We will briefly discuss the thermal excitation of spin waves or magnons.

We consider a linear chain of N spins, all of which are oriented parallel to each other. According to (10.20) the total energy is

$$U = -2J \sum_{p=1}^{N} S_p S_{p+1}. \tag{10.28}$$

From this we obtain the energy of the ground state

$$U_O = -2NJS^2. \tag{10.29}$$

As a possible excitation, we look at the case, where a single spin is oriented antiparallel to its neighbors (Fig. 10.5b). The energy increase ΔU amounts to

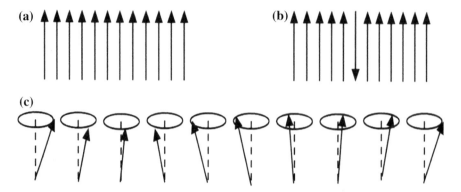

Fig. 10.5 Elementary magnets in a simple ferromagnet. **a** In the ground state all elementary magnets are oriented in the same direction. **b** A possible excitation from the ground state: one spin is flipped over. **c** Spin wave in a chain of spins in a perspective presentation

$$\Delta U = 2\,J\left(2S^2 + 2S^2\right) = 8\,JS^2, \tag{10.30}$$

where in the bracket we have separately listed the interaction with the left and the right neighbor. However, a much smaller excitation energy is needed, if the change in the spin orientation occurs only gradually. Exactly this is accomplished by the excitation of spin waves (Fig. 10.5c). In the case of the ferromagnetic magnons one obtains the dispersion relation from the equation for the temporal change of the angular momenta of the spins. In the limit of small excitation amplitudes, one finds the dispersion relation

$$\hbar\omega = 4\,JS(1 - \cos\,ka), \tag{10.31}$$

where a denotes the lattice constant of the chain and k the wave number. In the limit of large wave lengths (ka $\ll$ 1), from (10.31) one obtains

$$\hbar\omega = 2\,J\,S\,a^2k^2. \tag{10.32}$$

We note, that the dependence $\omega \sim k^2$ is different than in the case of phonons with $\omega \sim K$ (see (3.5) and (3.6)). These results can be easily extended to the three-dimensional crystal lattice.

In the case of the thermally excited magnons the Bose-Einstein distribution (3.2) is valid. The energy of the magnons is

$$U = \int d\omega\,D(\omega)\langle n_\omega\rangle\hbar\omega \tag{10.33}$$

analogous to (3.3), where the integral is taken over the first Brillouin zone. In the limit of low temperatures (ka $\ll$ 1), for the density of states one finds $D(\omega) \sim \omega^{1/2}$, yielding $U \sim T^{5/2}$. The resulting specific heat of the magnons is

$$C_V = \left(\frac{\partial U}{\partial T}\right)_V \sim T^{3/2}. \qquad (10.34)$$

Because of the thermal excitation of magnons, the magnetization is reduced by the amount $\Delta M = M(0) - M(T)$. The final result is

$$\Delta M / M(0) \sim T^{3/2}. \qquad (10.35)$$

We see that the thermal excitation of spin waves or magnons affects the specific heat and the saturation magnetization of a ferromagnet. In both cases the contribution of the magnons leads to the famous $T^{3/2}$ law derived by Felix Bloch for the temperature dependence of the quantities.

Magnons contribute also to the heat conductivity in crystals, and they influence the electrical transport properties such as the electrical conductivity and the thermoelectric phenomena. Similar to the case for phonons, the energy spectra of the magnons can also be determined experimentally by means of inelastic neutron scattering.

10.5   Antiferromagnetism

In addition to the parallel orientation of the spin magnetic moments of ferromagnetism, there also exists the case where the spins of the neighboring atoms in the crystal are oriented exactly antiparallel to each other. This case is referred to as antiferromagnetism. For antiferromagnetism the "quantum mechanical exchange integral" (10.21) is negative, whereas for ferromagnetism it is positive. Already in the late 1920s, the Frenchman Louis Eugene Felix Néel had the idea that there must also exist another kind of magnetic order in crystals, in which the spin magnetic moments of neighboring atoms are oriented exactly antiparallel to each other. After completing his studies in Paris, in the year 1928, Néel took the position of assistant of Pierre Weiss in Strassburg. Néel devoted his whole career to magnetism, up to 1940 in Strassburg and subsequently in Grenoble. Later on, largely due to Néel, has Grenoble developed into the important center of Materials Science and Solid State Physics in France, for which it is today well known everywhere. In his early idea of antiparallel spin orientation of neighboring atoms in the crystal, Néel had assumed that two lattices were penetrating each other, each of these "sublattices" by itself showing ferromagnetic order, but both being magnetized exactly in opposite direction to each other. Hence, overall the crystal remains magnetically neutral.

Therefore, the experimental evidence for such a novel possibility of magnetic order contemplated by Néel was difficult to obtain. In 1938 measurements with manganeseoxide (MnO) yielded the first positive results. The final confirmation of the hypothesis of the two sublattices penetrating each other, which are magnetized in opposite directions to each other, was achieved in the year 1949 by means of elastic neutron diffraction experiments. The antiferromagnetic order vanishes above the "Néel temperature". The notation antiferromagnetism was proposed in 1938 by the American Francis Bitter in a theoretical paper. A famous example, much discussed in recent years, is the antiferromagnetic order of the spin magnetic moments of copper atoms in the copper-oxide planes in the undoped state of materials showing high-temperature superconductivity after doping.

In an antiferromagnetic material, at finite temperatures, antiferromagnetic spin waves are thermally excited. One refers to antiferromagnetic magnons. Antiferromagnetic spin waves show the dispersion relation

$$\hbar\omega = 4|J||S||\sin ka| \tag{10.36}$$

similar to the case of phonons (see (3.5)). In the limit $ka \ll 1$ we have

$$\hbar\omega = 4|J||S||ka|. \tag{10.37}$$

Antiferromagnetic spin waves contribute to the specific heat and to the heat conductivity of the crystals, with a temperature dependence proportional to T^3 at low temperatures, similar to phonons. Again, their energy spectrum can be determined experimentally by means of inelastic neutron scattering.

In addition to the two discussed magnetic ordering phenomena of ferromagnetism and antiferromagnetism, there also exist other forms of ordered magnetic structures. However, we will not discuss these in further detail.

In this chapter our whole discussion of magnetism has been limited to the electrons acting as elementary magnets. However, there also exists nuclear magnetism associated with the atomic nuclei and their quality as elementary magnets. Since the elementary magnets of atomic nuclei are about two thousand times weaker than those of electrons, the effects of nuclear magnetism are restricted only to very low temperatures. For example, in the paramagnetism of atomic nuclei, the ratio between the magnetic field and the temperature must be two thousand times larger compared with the paramagnetism of electrons, in order to achieve the same degree of alignment of the nuclear spins. Therefore, we will refrain from any further discussion of nuclear magnetism.

10.6   Technical Applications, Giant Magneto-Resistance, Spintronics

For a long time ferromagnetic materials have been interesting for their technical applications. In many offices and homes we find small sticking magnets. There are of course more ambitious applications of permanent magnets in electric and in transportation technology.

Often magnetic couplings provide distinct advantages. The magnetic "hardness" of a material for a permanent magnet plays an important role in their applications. One distinguishes between magnetic soft and magnetic hard materials. The magnetic hardness is quantified in terms of the "coercive field" H_C. The latter indicates the value H_C of the magnetic field, at which the unmagnetized state of the material is again reestablished, if this magnetic field is applied a second time in a direction opposite to that of the original magnetization of the material. This is shown schematically in Fig. 10.6, where the magnetic flux density B is plotted versus the applied magnetic field H in the case of a magnetic soft and of a magnetic hard material. A soft material shows a small value of H_C and little hysteresis, in contrast to a magnetic hard material with a large value of H_C and large hysteresis.

In this context we remember that, in all individual magnetic domains we have discussed above, at first the magnetization must be rotated into the same direction by means of a suitably applied magnetic field, in order to obtain a strong permanent magnet. Alloys consisting of aluminum, nickel, and cobalt (AlNiCo) belong to the oldest and mostly tried materials for permanent magnets. Record values of the coercive field are achieved in alloys of samarium and cobalt ($SmCo_5$). Sinter materials such as, for example, barium- or strontium-ferrite, fabricated from a magnetic powder of small single-domain particles, are economically highly attractive. For the large-scale technical project of the magnetic suspension train "Transrapid" the latter sinter materials are particularly suitable as levitation magnets.

Fig. 10.6 Magnetic flux density B plotted versus the applied magnetic field H in the case of a magnetic soft (left) and a magnetic hard (right) material

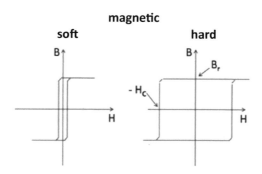

H_C = coercive field
B_r = remanence

In recent years magnetism has increasingly entered the field of microelectronics, where it has triggered very interesting developments. In this case, in addition to the electric charge, the spin and the associated magnetic moment of the electron play a significant role in magneto-electronic devices. The large advances in technology for the preparation of thin layers and multi-layer packages have been an important prerequisite for this development. Here nearly atomic accuracy in the fabrication of the layers has been achieved. Today this field is referred to as magneto-electronics, spin-electronics, or spintronics. While usually in electronic circuits the spins of the electrons are arbitrarily oriented and do not influence the electric current flow, in spintronics "spin-polarized" electric currents are used, where the spin of the mobile electrons is oriented in a specific direction. In this case the spin serves to control the electric current flow. Important fields for the application of magneto-electronics exist in magnetic technology for data handling in computers, for example, in the reading heads for hard disks and in the magnetic elements for data storage. In addition, we mention the magneto-sensorics in automotive technology, mechanical engineering, and in medicine. At present a highly promising goal for the not too distant future is a close combination of magneto-electronics with semiconductor technology.

In the previous generation of magnetic sensors, for example, in the reading heads for extracting data stored in hard disks, the electrical resistance change of a ferromagnetic layer due to an external magnetic field, is utilized. The magnetic data storage is based on small magnetic domains, which represent the "0" or "1" of the digital information by means of their different magnetization. In the reading head the electrical resistance change serves to detect the local magnetic field at the surface of the hard disk and thus the digital information. In the year 1988, this technology experienced an important advance, when Peter Grünberg at the German Research Center in Jülich and, nearly simultaneously, Albert Fert at the Université Paris Sud discovered the "giant magneto-resistance". Two years prior to this discovery, Peter Grünberg had observed an unusual magnetic behavior in a multi-layer package consisting of iron and chromium. Apparently, there is a coupling between two ferromagnetic layers of iron, which are separated from each other by a thin, nonmagnetic and metallic layer of chromium, such that the magnetization of neighboring iron layers is oriented either parallel or antiparallel. Which of these two kinds of couplings occurs, depends on the thickness of the nonmagnetic layer in-between, and varies between antiparallel and parallel with increasing thickness of the layer. Now, during electric current flow along the package of the layers, the electrical resistance depends sensitively on whether two neighboring iron layers are magnetized in the same or the opposite direction. (From a simple energy argument we would expect, that two neighboring iron layers would be magnetized in mutually opposite direction, because in this case the magnetic fields of the return flux would cancel each other between both layers, and the magnetic field-energy would be reduced correspondingly).

For our further discussion we assume a multi-layer package, consisting of several ferromagnetic iron layers, where two neighboring iron layers are separated from each other by a thin nonmagnetic chromium layer, respectively. In the absence

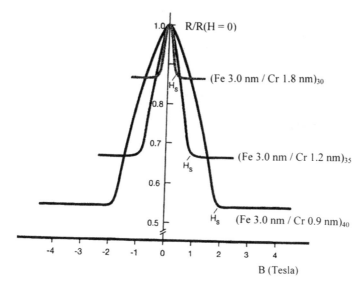

Fig. 10.7 Giant magneto-resistance: Electrical resistance (in units of the resistance at zero magnetic field) plotted as a function of the external magnetic field B for different iron-chromium multi-layers at the temperature of 4.2 K. H_S denotes the magnetic field, at which the magnetizations of the iron layers become oriented in parallel. Identification of the multi-layer structures: for example, (Fe 3.0 nm/Cr 1.2 nm)$_{35}$ denotes a multi-layer package consisting of 35 double layers of one 3.0 nm layer of iron and one 1.2 nm layer of chromium. (M. N. Baibich)

of a magnetic field, two neighboring iron layers are assumed to be magnetized in opposite direction with respect to each other. In this case of "antiferromagnetic coupling" between the layers the electric current flow along the package is hindered because of a relatively large resistance. If an external magnetic field is applied parallel to the multi-layer package, in all iron layers the magnetization orients itself along the direction of the magnetic field, and the electrical resistance shows a strong decrease with increasing magnetic field. This is the basic principle of giant magneto-resistance (Fig. 10.7). Such a multi-layer arrangement can still be generalized by dropping the requirement of the antiferromagnetic coupling between the layers. For example, one can imagine magnetic multi-layer systems, in which the magnetization in one ferromagnetic layer is fixed, whereas in the other layer the magnetization can be rotated back and forth. This can be accomplished by means of a large difference in the coercive field of the two ferromagnetic layers. In this case a relatively thick nonmagnetic layer in between is also possible. Such multi-layer systems showing giant magneto-resistance even without antiferromagnetic coupling between the layers are referred to as spin valves.

The giant magneto-resistance of the spin valves has technically been applied for some time in the reading heads for extracting the data stored in computer hard drives. Within 10 years of the discovery, this technical application has developed into a billion-dollar business. Whereas in one of the two ferromagnetic layers the direction of the magnetization is fixed, in the other layer the direction is freely adjustable. As the reading head glides along the surface of the hard disk, because of the small magnetic fields representing the stored digital information in the form of "0" or "1", the magnetization in this other ferromagnetic layer is rotated back and forth. Simultaneously, the electric current flow changes correspondingly, in this way yielding the output signal. Since still weaker magnetic fields can be detected by these reading heads, compared with their predecessors, the density of the data stored in the hard disk can be increased by about a factor of three.

The magnetic tunnel junction, consisting of three layers, represents another magneto-electronic device. In this case, two ferromagnetic metal layers are separated from each other by an electrically insulating metal-oxide layer with a thickness of only 1 nm. Electric current flow across the junction is only possible because of the quantum mechanical tunneling process. Similar to the spin valve discussed above, the electric tunneling current can flow without additional resistance only if both ferromagnetic layers are magnetized in the same direction. In the opposite case, the tunneling current experiences a high resistance. Again, the direction of the magnetization in one of the two ferromagnetic layers is fixed, whereas in the other layer it can be pointed in the parallel ("0") or in the antiparallel ("1") direction, and in this way it can be used for the storage of a unit of digital information. A program for the mass production of these "MRAMs" (magnetic random-access memories) for data storage, based on magnetic tunnel junctions, has been started jointly by the Companies IBM and Infineon. The industrial fabrication of 256-kbyte MRAM chips has been reported recently.

As the last example of magneto-electronic devices we discuss an interesting proposal of a field-effect transistor operating with spin-polarized electric currents (Fig. 10.8). In this case the electric current is carried by a two-dimensional electron gas at the interface of a semiconductor heterostructure of indium-galliumarsenide (InGaAs) and indium-aluminumarsenide (InAlAs). The two-dimensional electron gas at a semiconductor interface had already appeared in Chap. 7 in our discussion

Fig. 10.8 Schematics of a spin-polarized field-effect transistor. The electric current passes through the two-dimensional electron gas at the interface of a semiconductor heterostructure of InGaAs and InAlAs. The current flow is controlled by the voltage V_g applied to the gate electrode

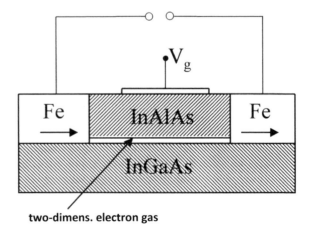

two-dimens. electron gas

of the integral and the fractional quantum Hall effect. In the present case, the two-dimensional electron gas represents a current channel with a very high mobility of the electrons. Furthermore, the channel is assumed to be free of collision processes, which can flip the spin of the moving electrons. At both ends of the channel, ferromagnetic metal contacts inject spin-polarized electrons into the channel or extract them again. At the top surface of the semiconductor heterostructure a metal electrode is attached, with which an electric gate voltage can be applied perpendicular to the current channel. The current flow into the magnetized metal contact acting as the collector depends sensitively upon the direction of the spin polarization of the incoming electrons. The electric current can flow almost without any hindrance only if the spins of the electrons are pointing in the same direction as the magnetization in the collector. Otherwise, the current experiences a relatively high resistance. However, in the electric field generated perpendicular to the current channel by means of the gate voltage, the spins of the rapidly traversing electrons are rotated. Hence, the electrical resistance of the transistor can be controlled and modulated by means of the gate voltage. Here the rotation of the spin orientation in the electric field directed perpendicular to the current channel, is caused by an effect which is explained only by the theory of relativity and which we do not pursue any further here.

The examples discussed clearly demonstrate the high potential of magneto-electronics for technical development, which is by no means yet exhausted. Today the most important memory systems for data storage are based on magnetic devices. It is interesting to recall again the impressive development of storage density on hard disks. In a little more than 40 years, from 1956 until 2000, storage density increased by a factor of about ten million. In the year 2000 it was about 2.6 GB per cm^2, and in the year 2011 it reached about 65 GB per cm^2 (GB = 10^9 bits). Another important advantage of spintronics results from the fact that the rotation process of the electron spin consumes only very little energy and occurs extremely quickly.

Chapter 11
Nanostructures: Superlattices, Quantum Wires, and Quantum Dots

Abstract The advanced microfabrication techniques can produce objects which are sufficiently small, that new quantum effects appear. After discussing superlattices and Bloch oscillations, we turn to the Landauer transmission channels and the quantized conductance of quantum wires. The fullerenes are carbon molecules consisting of different distinct numbers of carbon atoms. Their outgrowth in form of carbon nanotubes represents a promising structure for molecular electronics. Conducting layers of graphene with a thickness of only a single carbon atom display a fascinating electronic band structure, where the concentration of charge carriers can be varied strongly by means of an attached gate electrode. Quantum dots find increasing use in optoelectronics. Topological insulators show surprising current-carrying edge or surface states. The topological properties of the electronic band structure and possible contact points between the valence band and the conduction band and band crossings turned into important points of modern developments.

In December of 1959 Richard P. Feynman, one of the most brilliant American physicists of the last century, presented a visionary and highly acclaimed lecture with the title "There is Plenty of Room at the Bottom". At that time Feynman had already foreseen something that would be confirmed impressively during the following decades because of the rapidly advancing miniaturization in the field of microelectronics. He derived one of his leading ideas from the perception of molecular biology at that time, that only about 50 atoms within the DNA double helix are needed for one bit of biological information. The winter of 1952/1953, when Rosalind Franklin of Kings College in London had, with her X-ray images, confirmed for the first time the double helix structure of DNA, was still relatively close. If, for comparison, we assume a geometric structure size of an electronic device of 45 nm, the limit which can just be reached by 2012, we find for the total number of crystal atoms within a little cube having sides of length 45 nm, the huge number of one million atoms. Here we have taken an average distance of 0.5 nm between the atoms in the crystal. From this comparison we can clearly see, how

R. P. Huebener, *Conductors, Semiconductors, Superconductors*, Undergraduate Lecture Notes in Physics, https://doi.org/10.1007/978-3-030-31420-0_11

much room there still is today "at the bottom", compared with the molecular level of biology.

It has been the drive for continuous miniaturization in the field of microelectronics, which has provided the motivation for improvements in the methods of fabricating microstructured solid objects. In this context, great advances were achieved in the technique for the preparation of thin layers and of multi-layer packages of the relevant materials. Special tricks employed during the deposition on the substrate and during the subsequent etching process, made it possible to fabricate smaller and smaller objects from the thin layers. In the meantime lithography methods have been extended to ultraviolet light and X-rays, in order to achieve higher spatial resolution with much shorter wave lengths. Also, electron beams with high energy are utilized for lithography, and, recently, high-energy ion beams of helium and hydrogen ions have been tested for their suitability to achieve still smaller structural sizes. In many cases these fabrication processes must be carried out in ultra-high vacuum, and only ultra-pure materials can be used. The word "ultra" now appears more and more often in this field. Together with the technology for the fabrication of thin layers, the methods for controlling and analyzing the structure and composition of the layers have been continuously improved, and today reach nearly atomic accuracy, if needed. Layers and multi-layer packages of different materials stacked on top of each other can be fabricated, having a microscopically single-crystalline structure. Transmission electron microscopy and scanning probe microscopy allow the analysis of the materials and in particular of their surfaces with atomic spatial resolution. In the meantime, scanning probe microscopy has been applied successfully also at low temperatures. Finally, the techniques for micromanipulation have been developed further, such that it has become possible to perform electric and mechanical measurements even on single atoms and molecules.

During the continuously advancing miniaturization of fabricated objects, one finally reaches spatial scales at which new quantum effects appear. These effects always result because of the nature of the electrons acting as quantum mechanical matter waves. In the following we will illustrate this effect with a few examples.

11.1 Superlattices, Bloch Oscillations

As we discussed in Chap. 2, in a crystal the elementary cell repeats itself along all three spatial directions and in this way generates the three-dimensional periodic structure. This principle is also found in the case of the superlattices, in which the composition of a material is periodically modulated along one spatial direction (Fig. 11.1). Such multi-layer structures are fabricated by means of modern thin-film technology.

In the year 1970 Leo Esaki and Ray Tsu started to think about and to fabricate "superlattices" out of semiconductors. At the time both worked at the American Thomas J. Watson Research Center of IBM in Yorktown Heights in the Federal State

Fig. 11.1 In a superlattice, thin layers of two different metals or semiconductors are alternately stacked on top of each other with high regularity

of New York. Already by the late 1950s Esaki had gained much attention because of his research dealing with the electrical behavior of the "Esaki diode". Then he had studied the unusual features of the electrical resistance of p-n junctions in semiconductors and had identified the quantum mechanical tunneling process as the crucial underlying mechanism. The quantum mechanical tunneling process allows a particle to pass through a relatively high energy barrier, which classically would be impossible. This process would also play a central role in the Esaki superlattices made from semiconductors. Such a superlattice is fabricated by alternately placing thin layers of two different metals or semiconductors on top of each other during the deposition process. Here great attention must be paid to the atomic accuracy of each layer and to the perfect periodicity of the spatial sequence of the layers. For his experiments Esaki utilized superlattices fabricated from the two semiconductors galliumarsenide and aluminum-galliumarsenide (GaAs/AlGaAs), since this combination of materials yielded samples with the highest quality. As we have discussed in Chap. 7, later on the same semiconductor system GaAs/AlGaAs was used for the fabrication of the two-dimensional electron gas, in which the fractional quantum Hall effect was discovered. In his superlattices, Esaki has stacked up to 100 double layers of GaAs and AlGaAs on top of each other. In such a superlattice the length of a spatial period is about 10 nm, and hence this length is 20–40 times larger than the distance between the atoms in a typical crystal lattice. Here the superlattice structure only exists along one direction, namely perpendicular to the planes of the individual layers.

Whereas the quantum mechanical wave function of the electrons does not change very much along the directions parallel to the planes of the layers in the superlattice, along the perpendicular direction the periodicity of the superlattice has a strong influence. Exactly in the same way as electrons, in the form of matter waves, experience Bragg reflection at the crystal lattice, leading to the gaps in the energy spectrum of the electrons, so Bragg reflection also occurs at the periodic structure of the superlattice, and new energy gaps appear. In Chap. 4 we have treated the appearance of Bragg reflection (Fig. 4.4), when the wave vector **k** of the electrons reaches the boundary of a Brillouin zone and the forbidden gaps in the energy spectrum (Figs. 4.3 and 4.5).

In the example shown in Fig. 2.9 Bragg reflection occurs at the values $k_x = \pi/a$ and $k_y = \pi/b$ of the wave vector **k**, where a and b denote the lattice constants along the x- and y-direction, respectively. In a superlattice, the (super-) lattice constant along the direction of the modulation (perpendicular to the planes of the layers) is much larger than the lattice constant of the underlying crystal. The wave number, at which the Bragg reflection takes place, is inversely proportional to the distance between two neighboring lattice points within the underlying spatially periodic lattice structure. Therefore, in superlattices Bragg reflection is expected already at correspondingly much smaller values of the wave vector compared with the case of the crystal lattice. As a result, along the directions perpendicular to the layers of the superlattice, there exist energy bands which are much narrower than the usual energy bands, and which are referred to as "minibands". The existence of these minibands leads to important changes in the electrical properties of the semiconductor superlattice. Ultimately, this was the reason why Esaki studied his superlattices at the time. He was hoping that new and particularly fast electric oscillators would be discovered.

If an electric voltage is applied to the superlattice parallel to the direction of the modulation, in the relevant miniband the electrons are accelerated in the direction of the current flow, and they experience a gain in their energy. However, because of the very narrow energy width of the miniband, there is the possibility that the electrons will reach the upper edge of the miniband without first losing some energy in a collision process (Fig. 11.2). At the upper edge of the miniband the electrons are reflected, since they cannot traverse the adjoining energy gap in order to reach the next higher miniband. This is exactly the process of Bragg reflection experienced by the electrons representing quantum mechanical matter waves.

We illustrate this with the example of a one-dimensional periodic chain of atoms with the lattice constant a. The energy spectrum between the boundaries of the first Brillouin zone, π/a und $-\pi/a$, is shown in Fig. 11.3. We start from (5.1) for the force acting on the electrons and obtain

$$\left|\frac{\Delta k}{\Delta t}\right| = e\,E/\hbar. \tag{11.1}$$

In the absence of any scattering processes, in an electric field, an electron (or hole) gains momentum and energy until the value $k = \pi/a$ is reached. By means of a Bragg reflection, the electron is reflected from $k = \pi/a$ to $k = -\pi/a$, and subsequently its wave number increases again due to (11.1). This process repeats itself and results in a periodic oscillation of the electrons, the "Bloch oscillation". The time needed for increasing the wave number by the amount $\Delta k = 2\pi/a$, we denote by τ_B. From (11.1) we obtain

$$2\pi/a = e\,E\,\tau_B/\hbar \tag{11.2}$$

and

Conduction Band

Miniband

Δz

Phonon Emission

Fig. 11.2 In an electric field, the energy bands are inclined in the field direction because of the potential gradient, and during their motion the electrons are brought closer to the upper edge of the conduction band. However, in a standard semiconductor (left side) the electrons already experience a collision process, during which their energy is lowered because of the emission of a phonon, long before they reach the upper edge of the band. On the other hand, in the narrow miniband of a semiconductor superlattice (middle) the electrons reach the upper band edge long before a collision process takes place. At the upper band edge they undergo a Bragg reflection, and this process repeats itself as a "Bloch oscillation", until it is interrupted by a collision process. The enlarged section of a miniband (right side) shows how the electrons move by a distance Δz in the field direction with each collision process, at which their energy is lowered by the emission of a phonon

Fig. 11.3 Energy spectrum $\varepsilon(k)$ in the first Brillouin zone between $k = \pi/a$ and $k = -\pi/a$

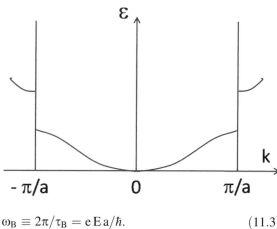

$$\omega_B \equiv 2\pi/\tau_B = e\,E\,a/\hbar. \qquad (11.3)$$

The angular frequency ω_B is the Bloch frequency. For the Bloch oscillation to occur, the average scattering time τ of the electrons must be sufficiently long and must satisfy the condition

$$\omega_B \tau \gg 1. \qquad (11.4)$$

Furthermore, for the appearance of the oscillations very pure materials and low temperatures are required. An additional important requirement is a large value of the lattice constant a, resulting in a small energy width of the miniband and in a correspondingly large value of ω_B according to (11.3).

Above all, it is the relatively small energy width of the minibands, which plays a crucial role in this behavior. Compared with a semiconductor superlattice, in a usual semiconductor crystal, the width of the energy bands is much larger. Therefore, in this case, during their energy gain in the electric field, the electrons already undergo a collision process due to the lattice vibrations long before they have reached the upper edge of the band. During such a collision process energy is always transferred from the electrons to the crystal lattice, and, hence, the upper edge of the energy band remains far away, and the Bloch oscillations cannot occur. However, for the minibands of the superlattice this is totally different (Fig. 11.2).

If the electric voltage applied perpendicular to the layers of the superlattice is increased more and more, eventually the electric potential difference between two neighboring unit cells of the superlattice becomes sufficiently strong that the cells are decoupled from each other. Whereas at relatively small electric fields the quantum mechanical wave function of the electrons extends spatially coherently over many cells of the superlattice, and the electronic structure of the minibands still remains intact, at high electric fields the wave function becomes more and more spatially localized at each individual cell, and decoupling between the cells occurs. Instead of a miniband extending over the whole superlattice, now in each cell individual discrete energy levels exist, which are adjusted to the electrical potential gradient across the superlattice. This splitting of the continuous energy of the miniband into the discrete energy levels is known as the "Wannier-Stark ladder". The name Wannier-Stark ladder originates from two distinguished physicists: in the beginning of the last century the German Johannes Stark had discovered the splitting of spectral lines due to an electric field, referred to since as the Stark effect, while the American Gregory Hugh Wannier had contributed significantly to the theoretical foundations of solid state physics.

During Bloch oscillation as well as during the splitting of a miniband into the individual energy levels of the Wannier-Stark ladder, the mobile electrons become localized within only a few, and eventually only within a single cell, of the superlattice, because of the electric field. This effect increases with increasing electric field, such that above a specific value of the electric field the flow of the electric current decreases with increasing voltage. In this case we have negative differential resistance. Instead of consuming energy, the superlattice can now return energy into an oscillating electrical circuit, in this way acting as an active device generating high-frequency electromagnetic waves.

The semiconductor superlattices studied by Esaki represent the case of the "heterostructure superlattices" fabricated layer by layer from two different semi-conductors. Esaki also had the idea that it should be possible to produce semi-conductor superlattices simply and with a high degree of flexibility, only with a single semiconductor, by means of an alternate spatially periodic n-doping and p-doping of this semiconductor. At the beginning of the 1970s, the German Gottfried H. Döhler, at first as a postdoctoral member of Esaki's group, took up this idea of the doping superlattice. Since the n- and the p-doped layers, respectively, are separated from each other by a thin, electrically insulating semiconductor layer, these superlattices are also referred to as "n-i-p-i crystals". The first n-i-p-i structures were fabricated in 1980 by Klaus Ploog at the Max Planck Institute for Solid State Research in Stuttgart. These experiments were performed with the semiconductor galliumarsenide (GaAs). Silicon atoms were used for n-doping and beryllium atoms for p-doping. Subsequent electrical and optical measurements with these doping superlattices have well confirmed their expected physical properties.

The possibility of fabricating superlattices from semiconductors has introduced an interesting additional option for the development of new materials in electronics and optoelectronics. In superlattices, the electrical and the optical properties can be varied artificially. In the meantime many experiments have been performed with semiconductor superlattices. However, at present, the technical applications are only in their very early stages. Highly promising developments concentrate on the "quantum cascade laser" operating in the infrared spectral range. Here transitions between the discrete energy levels are utilized, for which the emitted frequency can be tuned by the variation of the material composition and of the thickness of the layers. Recently, interesting progress has been reported also for the generation of microwaves by means of the Bloch oscillations of the electrons in semicondutor superlattices.

11.2   Mesoscopic Regime, Ballistic Electron Transport, Quantized Conductance Value

Eventually, the use of ultra-pure materials and the ability to fabricate objects with smaller and smaller dimensions made it possible, that within the studied sections of the electrically conducting materials, the electrons experience almost no collision processes, or only very rarely. The probability becomes extremely small that, in these experiments dealing with very small spatial dimensions, the measurements are influenced by many structural defects or chemical impurities in the crystal. Furthermore, at sufficiently low temperatures most of the lattice vibrations can be frozen out. Under these conditions, the spatial dimensions belong to the "mesoscopic regime", located between the single atoms or molecules on the one hand and the macroscopic world of events on the other hand. Within this mesoscopic length scale all aspects of the electrons as matter waves are fully valid, and the observed physical behavior of the electrons can best be understood in terms of a propagating wave (Fig. 11.4). In Sect. 5.2 we discussed the Fermi distribution of the energy of the electrons, resulting from the exact identity of electrons as elementary particles, and we have pointed out that, as a result, most electronic material properties are determined only by the electrons from the immediate proximity of the Fermi energy. Therefore, the unperturbed, "ballistic motion" of the electrons within the mesoscopic dimensions happens at the Fermi velocity. In the following we denote the Fermi velocity by v_F. Similarly to the Fermi wave number k_F, the Fermi velocity v_F is also fixed by the Fermi energy ε_F.

Fig. 11.4 Photograph, produced by an electron-transmission microscope, of a ring fabricated from a gold layer of 38-nanometer thickness. The inner diameter of the ring is 784 nm. The width of the conducting lines is 41 nm (R. A. Webb)

The unperturbed, ballistic propagation of particles or energy quanta is in contrast to the other limit, in which the propagation is always interrupted by collisions and deflections (Fig. 11.5). A well known example of the latter case is the propagation of light in dense fog, in which all contours disappear, and any orientation becomes impossible. This case is referred to as diffusive propagation, by means of the process of "diffusion". On the other hand, in the absence of fog we have ballistic and straight arrays of light, propagating with light velocity and clearly marking the spatial environment.

The electron transport in the mesoscopic regime in the form of a ballistically propagating wave is characterized by the fact that the internal material properties of the object are no longer decisive, and that, instead, the shape of the external boundary has a much stronger influence. Now the electrons experience collisions and deflections predominantly only at the boundary of the object, for example at the entry or the exit of a constriction. The behavior is much more similar to that in a wave guide. Now the property of the electrons as a quantum mechanical matter wave dominates. Therefore, in the case of such conductors one also speaks of quantum wires. It was the American Rolf Landauer, who considered these questions for the first time in the year 1957. Landauer, who originally came from Germany, worked at the American Thomas J. Watson Research Center of IBM. At that time he proposed his famous concept of the transmission channels in mesoscopic electrical conductors, which subsequently turned out to be extremely productive and successful. We will briefly discuss his central idea. The quantum wire is assumed to be placed along the x-direction. At both ends it is connected to electrodes with the energy $\varepsilon + \frac{1}{2}$ eV and $\varepsilon - \frac{1}{2}$ eV, respectively. Here V is the electrical potential

Fig. 11.5 If the dimensions of an electrical conductor become smaller and smaller (in the figure from the top to the bottom), the collision processes within the interior of the conductor become less and less important, and the shape of the external boundary will eventually become crucial. In this case the electrons move ballistically as matter waves

diffusive

quasi-ballistic

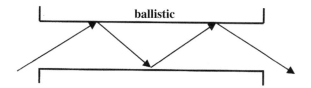

ballistic

difference between both electrodes. If we treat both spins of the electrons separately, we find for the current

$$I = -2e \int \frac{dk_x}{2\pi} v_x \left[f\left(\varepsilon + \frac{1}{2}eV, \varepsilon_F\right) - f\left(\varepsilon - \frac{1}{2}eV, \varepsilon_F\right) \right] T_{trm}. \qquad (11.5)$$

Here, for the Fermi distribution function (5.10), we have used the abbreviation

$$f(\varepsilon, \varepsilon_F) \equiv \frac{1}{e^{(\varepsilon - \varepsilon_F)/k_B T} + 1} \qquad (11.6)$$

and we have taken into account the current flow in both directions (+x and −x). The quantity T_{trm} is a transmission coefficient describing the scattering processes of the electrons, which establish the equilibrium with the local electrochemical potential of the electrodes. We assume that no scattering processes occur within the quantum wire. In the case of small voltage V and low temperatures we have

$$\left[f\left(\varepsilon + \frac{1}{2}eV, \varepsilon_F\right) - f\left(\varepsilon - \frac{1}{2}eV, \varepsilon_F\right) \right] = \left(\frac{\partial f}{\partial \varepsilon}\right)_{V=0} eV \qquad (11.7a)$$

$$= -\delta(\varepsilon - \varepsilon_F)eV. \qquad (11.7.b)$$

$\delta(x)$ denotes the Dirac delta funktion. With $v_X\, dk_X = (1/\hbar)d\varepsilon$ finally we obtain the conductance G:

$$G = \frac{I}{V} = \frac{2e^2}{h}\,T_{trm} = 2G_O T_{trm}. \tag{11.8}$$

The quantity $G_O = e^2/h$ is the quantized unit of the conductance.

Landauer had found, that the electrical conductance of a one-dimensional channel connecting two charge reservoirs has to be measured in quantized units of $2e^2/h$ (The conductance is defined as the inverse of the electrical resistance). In this context we remember the unit h/e^2 of the quantized Hall resistance, which we discussed in Chap. 7. The factor 2 of the quantized conductance, according to Landauer, originates from the fact, that here we discuss the case without a magnetic field and that, therefore, both spin orientations of the electrons contribute in the same way to the result. On the other hand, the quantized Hall resistance only appears in high magnetic fields, where both spin orientations clearly must be treated separately.

The first experiments dealing with ballistic electron transport across a spatial constriction in the mesoscopic regime were carried out in 1965 by the Russian Yurii Vasil'evich Sharvin. He worked in the famous Institute for Physics Problems in Moscow, which is also named the Kapitza Institute after its founder. In his experiments Sharvin used "point contacts", prepared by pressing the sharp tip of a metal needle onto the surface of a metallic single crystal. During the electric current flow, at low temperatures, he measured the electrical resistance of this arrangement. However, in these experiments with metals, the role of the electrons as quantum mechanical matter waves is not yet highly pronounced, since, because of the typically relatively large Fermi energy of the electrons in metals, the wave length is only about 0.5 nm and, hence, it is much smaller than the opening of the point contact. Then in 1988 important progress was reported, when almost simultaneously two groups discovered the quantization of the electrical conductance of specially structured semiconductor heterostructures fabricated from galliumarsenide (GaAs) and aluminum-galliumarsenide ($Al_x Ga_{1-x}As$). One group belonged to the University of Delft and to the Philips Research Laboratories in Eindhoven and in Redhill, the other group worked at the Cavendish Laboratory of the University of Cambridge. The semiconductor heterostructure used by both groups was very similar to that in which, a few years earlier, Tsui and Störmer had discovered the fractional quantum Hall effect. In the two-dimensional electron gas of the semiconductor heterostructure the Fermi energy is much smaller than in metals, and correspondingly the Fermi wave length of the electrons is about one hundred times larger than in metals. This provides an excellent opportunity to observe new quantum effects during the passage of electrons through a narrow opening.

The two groups used the following technique for the fabrication of the narrow one-dimensional channel between two wide charge reservoirs within the two-dimensional electron gas of the semiconductor heterostructure. They attached two correctly structured metal electrodes, acting as gate electrodes, to the top surface of the heterostructure. At the narrowest location the opening between the two gate electrodes was only 250 or 500 nm wide. By applying a suitably selected gate voltage, the sample regions directly below the gate electrodes can be emptied completely of the charge carriers, such that only a conducting channel with an opening width of 250 nm or 500 nm, respectively, remains between both gates. By further increasing the gate voltage, the channel can be constricted even more, until eventually the two wide charge reservoirs are completely separated from each other. During their experiments both groups found that the conductance of their one-dimensional channel shows a regular step structure as a function of the gate voltage, and that the individual plateaus of the steps appeared at integer multiples of the quantized unit $2e^2/h$ of the conductance (Fig. 11.6). These measurements were carried out at low temperatures below 1 K. Apparently, the variation of the gate voltage causes a continuous change in the channel width, such that the number of the discrete and quantized conductance channels increases with increasing channel width.

The experimental observation of the quantized conductance of a narrow mesoscopic channel can be looked at as a special case of the concept of transmission channels introduced by Landauer. In the meantime, many papers have appeared dealing with details of this novel quantization phenomenon. However, here we refrain from any further discussion.

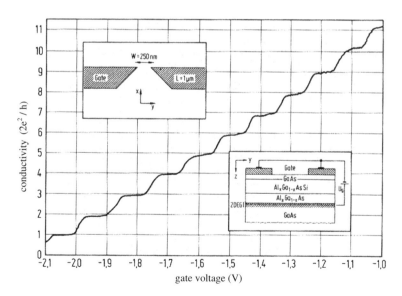

Fig. 11.6 Electrical conductance of a narrow one-dimensional channel in a GaAs/Al$_x$Ga$_{1-x}$As– heterostructure in the quantized unit ($2e^2/h$) plotted as a function of the gate voltage at a temperature of about 1 K. Inset on upper left: Arrangement of the gate electrodes on the surface of the heterostructure. Inset on lower right: cut through the heterostructure (2DEG: two-dimensional electron gas; U$_G$: gate voltage) (B. J. van Wees)

As the ultimate reduction in the size of an electrical contact between two charge reservoirs, in recent years even single atoms have been studied experimentally and theoretically. These experiments started at the French Commissariat à l'Energie Atomique in Saclay in the group of D. Esteve and Michel H. Devoret with the collaboration of the German guest scientist Elke Scheer. For measurements with individual atoms the technique of piezoelectric actuators, well-known from scanning tunneling microscopy, was employed, in addition to a special technique: the break junction method. The latter method allowed the mechanical control of the contact with an exceptionally high sensitivity. A suspended microbridge of about 2 μm length, 200 nm thickness, and with a 100 nm × 100 nm constriction in the middle, was mechanically stretched and eventually broken at the constriction. This was achieved by mounting the microbridge onto an elastic substrate, which could be bent mechanically in a highly controlled way. The authors studied atoms of different metals, such as lead, aluminum, niobium, gold, and sodium. Mostly, the experiments were carried out at temperatures much below 1 K. It was found that, with increasing stretching of the microbridges, the electrical conductance of the samples decreased in steps, until the contact was interrupted. The height of the individual steps was about $2e^2/h$. The quantized unit of the conductance again appeared. Landauer's concept of the transmission channels therefore also appears to be confirmed in this case. Furthermore, the experiments performed with the atoms of the different metals suggest that the number of conductance channels is equal to or at least closely related to the number of orbitals of the valence electrons of the central atom. For a quantitative understanding of electrical conductance properties of these contacts, a microscopic model must be developed, which takes into account the orbital structure of the atom as well as the local atomic geometry of the immediate environment. Electrical currents up to about 0.1 mA can pass through a contact consisting only of a single atom. This corresponds to the giant local electric current density of one hundred billion amperes per cm^2.

11.3 Bottom up, Fullerenes

For the nanostructures we have just discussed, the sample dimensions are reduced more and more, until eventually one reaches the mesoscopic regime where novel quantum effects in the electron motion can be observed. This method of operation is generally known as the "top-down" procedure. However, for the development of smaller and smaller devices aiming at "nanoelectronics" or eventually also at molecular electronics, the inverse procedure referred to as "bottom-up" gains much more importance. In this case, above all it is the methods of chemistry which are crucial for further advances. From this field of the molecular electronics, which presently shows a dramatic development, we wish to select a particular example: the "carbon nanotubes". However, first we must briefly illustrate the preceding history, which has lead to this spectacular development.

The physics and chemistry of the new forms of carbon started in astrophysics with the exploration of matter within interstellar space. During their experimental attempts to produce interstellar carbon molecules in the laboratory by means of laser evaporation of graphite, in 1985, Richard E. Smalley and Robert Floyd Curl at the American Rice University in Houston, Texas, and also Harold Walter Kroto at the University of Sussex in England, together with their co-workers, discovered the two carbon molecules C_{60} and C_{70} by means of mass-spectrometric analyses. At that time they had already presumed, that the C_{60} molecule possesses the structure of a soccer ball ("bucky ball"), in which the 60 carbon atoms are located at the corners of the five-cornered and of the six-cornered carbon rings, forming the nearly-spherical surface of the molecule. Altogether, the C_{60} molecule consists of 12 five-cornered and of 20 six-cornered carbon rings (Fig. 11.7). The discoverers called the molecule buckminster-fullerene after the American architect Buckminster Fuller, who was famous because of his buildings with a domed structure. Also the C_{70} molecule is composed of 12 five-cornered carbon rings, but 25 six-cornered carbon rings. It is stretched slightly and looks similar to an American foot ball. All carbon molecules with an all-round completely closed structure are now denoted as "fullerenes". Incidentally, more than 200 years ago the Swiss mathematician Leonhard Euler had already proved, that all fullerene structures must have exactly 12 five-cornered rings, in order to have an all-round completely closed shape.

Smalley and Kroto could only produce their fullerene molecules in such tiny amounts, that many supplementary studies and in particular crystallographic structure analyses were impossible. In the year 1990 this changed abruptly, when Walter Krätschmer of the Max Planck Institute for Nuclear Physics in Heidelberg and Donald R. Huffman of the American University of Arizona in Tucson succeeded for the first time in producing fullerene molecules in much larger amounts

Fig. 11.7 Perspective representation of a soccer-ball-shaped C_{60} molecule

than was possible before. Again, both scientists were interested in the preparation of soot particles, because they were dealing with questions about interstellar matter. In their preparation technique they used two rod-shaped graphite electrodes, between which an electric arc is burning with a high current density. During this process the electrode material evaporates. The whole preparation is carried out in an evaporation system, its recipient being filled with a cooling gas (typically helium). Because of the presence of the cooling gas, the carbon vapor condenses into a smoke of particles, which are then collected. The soot particles and the fullerene molecules are separated from each other by chemical methods. Since May 1990, Krätschmer and also Huffman were able to produce about 100 mg fullerene per day. Now the preparation of single crystals, of microcrystalline powder, and of thin layers followed quickly, and research activities started to grow explosively in many groups. In particular, the initial presumption of the soccer-ball structure of the fullerene molecules was exactly confirmed experimentally. The production method was improved by many groups, and it was scaled up for larger quantities. Now the experiments were extended also to solids consisting of C_{60} molecules, and electronic properties, as well as the influence of doping with admixtures, were investigated. Following the implantation of strong donors into the C_{60} solid (n-doping), even superconductivity was found with maximum values of the critical temperature up to $T_c = 48$ K. Here, the alkali metals potassium, rubidium, and cesium, as well as the alkaline earth metals, were mainly used for electron doping.

The C_{60} and the C_{70} molecules stand out because of their particularly high stability. Hence, during production their yield is also very high. However, the series of the fullerene molecules still extends much further. For example, there are the "magic" higher fullerenes C_{76}, C_{78}, C_{82}, and C_{84}. In the year 1991 Sumio Iijima from Japan, made a discovery with important consequences for the technical applications, when he observed in the electron microscope for the first time a new fullerene type in the shape of thin tubes like a needle. With this discovery of carbon nanotubes, a new phase in the fullerene research had begun. Iijima worked at the Laboratory for fundamental research of the Japanese NEC Corporation in Tsukuba. Since the discovery of carbon nanotubes, the number of publications and also the number of issued patents, dealing with the nanotubes, has grown from year to year. The number of walls of the tubes can vary. In his first publication Iijima had already reported tubes with up to seven walls. The tube diameter also varies correspondingly and falls into a range of about 4–30 nm. The typical tube length is about a few μm. Recently, scientists at the American Rensselaer Polytechnic Institute in Troy in the Federal State of New York and at the Chinese Tsinghua University in Beijing have prepared bundles of single-wall nanotubes even up to a length of 20 cm using a special technique (Fig. 11.8).

The electronic properties of the multi-wall nanotubes show relatively large variations, which severely hamper their reproducibility. In contrast to this, the single-wall nanotubes are very reproducible. Depending upon their diameter and upon the degree of angular rotation in their structural detail along the axial direction, referred to as the helicity, in terms of their electrical conductivity they behave like a one-dimensional metal or a semiconductor. Apparently, they are well suitable

Fig. 11.8 Single-wall carbon nanotube placed between two platinum electrodes. The width of the electrodes is 100 nm (C. Dekker)

for use as molecular wires. In addition to the fundamental physical properties of the carbon nanotubes, their potential for application in molecular electronics has been investigated by different groups. Here the manipulation of the nanotubes was accomplished using the methods of atomic force microscopy which we have mentioned in Chap. 1. Apparently, a sharp bend in the single-wall nanotubes acts like a rectifying diode, similarly as a metal-semiconductor contact. Such a bend can be generated by means of a pair of topological defects in the atomic structure of the nanotube, or by a local mechanical deformation. The function of a field-effect transistor has been demonstrated already by placing a single-wall nanotube on top of a gate electrode, electrically insulated from the tube. Last but not least, the carbon nanotubes can serve in an excellent way for realizing what are called single-electron effects in electrical transport properties. Here we mean that the physical properties of an object, such as the electrical resistance, for example, are strongly affected by the presence or absence of only a single electron, because of the extremely small dimensions of the object. Because of these developments, large companies within the computer industry presently show a keen interest in the physics and technology of carbon nanotubes. One can hear speculations already that, in the medium-range future, carbon nanotubes may start to compete with the comparatively expensive silicon as the substrate in the semiconductor technology. Carbon nanotubes have an extremely high conductivity for electric currents, and they allow densities of the electric current flow, at which copper wires would have melted long before. Hence, compared to conducting lines made from copper, the carbon nanotubes tolerate much higher electric power levels and operating frequencies. An additional interesting aspect of nanotubes arises due to the possibility that the tubes can be opened at both ends, and that molecules of other substances can be packed into their interior. In this way the tubes can be utilized as carriers of different materials.

11.4   Graphene

We imagine that the wall of the thinnest carbon nanotube is cut open along the axis, spread out and extended in both directions within the plane. Then we deal with graphene, i.e., with a material with a thickness of only a single carbon atom. The carbon atoms are arranged as a network with a honeycomb structure. In the year 2004, Andre Geim and Konstantin Novoselov succeeded in preparing a graphene sample by using a trick, and subsequently they discovered its fascinating physical properties. By using Scotch tape, they achieved to strip off the top atomic layers of a very pure graphite crystal and to transfer them onto a suitable substrate, such that finally individual flakes of one-atomic layers became available. Following their initial publications on this subject, an international competing research activity sat in. Within five years, from 2005 until 2009, more than 5000 papers about graphene were published.

Perhaps the most fascinating property of graphene is its electronic band structure. In the undoped state, it shows a linear increase of the energy E of the electrons with their wave number k: $E \sim k$. Therefore, the charge carriers act like relativistic particles without mass. Instead of the Schrödinger equation, they satisfy the Dirac equation, and the Fermi velocity v_F takes the role of the light velocity. This is in contrast to the usual quadratic increase $E \sim k^2$ in the case of a three-dimensional crystal (see (4.13) and Fig. 4.5a). In graphene, as the thinnest existing material, the concentration of the charge carriers can be varied strongly, by means of an attached gate electrode and the electric field effect. In this way the concentration of the electrons can be changed from that of a regular metal with about 10^{21} cm^{-3} to that of a metal with a similar concentration of holes, passing through the (semiconducting) state with only very few charge carriers (Fig. 11.9). In a magnetic field, Geim and Novoselov have also clearly observed the half-integer quantum-Hall-effect. The half-integer value arises from the fact that, because of the energy spectrum, at exactly $E = 0$ there are two energy levels, one with electron character and one with hole character. As an interesting exception, in graphene the quantum-Hall-effect could be observed even at room temperature.

In addition to its spectacular electronic properties, graphene has also exceptional mechanical properties. It is harder than diamond, and can become interesting for applications as composite material.

Motivated by the characteristic electronic band structure of graphene with its honeycomb-lattice-structure, very recently honeycomb-super-lattices were investigated experimentally and theoretically. Experimentally, these two-dimensional super-lattices consisting of HgTe nano-crystals were fabricated by Daniel Vanmaekelbergh and coworkers at the University of Utrecht using colloid-chemical methods.

In this case we deal with the most advanced stage of present research. Experiments on Bloch-oscillations and negative differential resistance due to the electronic mini-bands, as we have discussed in Sect. 11.1, still have to be carried out.

Fig. 11.9 In graphene, the concentration of the charge carriers and the electrical conductivity can be varied strongly, by means of an attached gate electrode and the electric field effect. The figure shows the dependence of the conductivity upon the gate voltage and the linear dependence of the Energy E of the electrons upon the wave vector **k**, E(**k**), for three values of the Fermi energy E_F, in the case of negative, vanishing, and positive gate voltage (A. Geim)

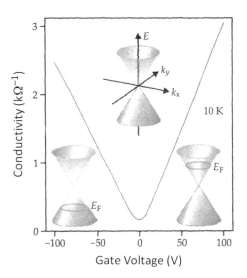

11.5   Quantum Dots

Carbon nanotubes have presented us with a highly promising molecular example as the smallest possible version of a quantum wire. Now we will also drop the last remaining dimension of the spatial extension of these one-dimensional quantum wires. This means that we are dealing with the "quantum dots" as objects with quasi-zero dimension. Again, it is the shape and dimension of the external boundary, which determine the physical behavior of the electrons within the quantum dots. On the other hand, the collision and deflection processes of the electrons in the interior of these objects are moving far into the background. Similar to the situation in an atom, now the quantum mechanical wave function of the electrons is determined to a large extent by the spatial size of the quantum dot. Hence, the quantum dots are also referred to as "artificial atoms". Energy bands for the electrons, such as those in an extended crystal, no longer exist. Instead of the energy bands, the electrons can occupy only discrete energy levels, which can be calculated from the geometric dimensions of the quantum dots using the quantum mechanical Schrödinger equation (This is schematically indicated in Fig. 7.3 at the bottom). In some sense, the Periodic Table of the Elements can be imitated by the occupation of individual energy levels of the quantum dots with electrons. Here the Pauli Principle must be obeyed by the electrons as Fermi particles. Hence, each state can be occupied by only two electrons, the spins of which are oriented in opposite directions. However, as an important difference between the quantum dots and the individual atoms we must note that the former are micro-crystals, consisting of about one thousand up to one million atoms, in which lattice vibrations (phonons) and also lattice defects exist. The energy spectrum of the electrons in the quantum dots can be found mainly from their optical properties, for example, from the spectroscopy of the energy transitions. It is also not surprising that, for technical applications of quantum dots, the optical properties are particularly interesting, as, for example, in the quantum-dot laser.

Quantum dots have been studied experimentally for about the last 20 years. For their fabrication three different general methods are employed. First we mention the relatively traditional "top-down" technique, in which the quantum-dot structures are defined and etched lithographically. However, the necessary processing steps are by no means simple. Furthermore, recently semiconductor nanoparticles, which were fabricated by methods from colloid chemistry, gained special importance. Above all, the II–VI compound semiconductors from the 2nd and the 6th group of the Periodic Table as well as the III–V semiconductors from the 3rd and the 5th group are interesting in the form of nanoparticles. The methods from colloid chemistry yield particles with quasi-spherical shape, the sizes of which can be produced reproducibly from only a few molecules up to highly extended dimensions. Here particles with a diameter between about 1 and 6 nm are particularly interesting, since their fabrication with other techniques is very difficult. This particle size falls into the regime of strong quantization, in which the distance between the discrete energy levels of the electrons in the quantum dots has the same order of magnitude as the energetic band gap in an extended crystal. As a rule, at the end of particle synthesis, a fractionating step for the separation of the particle sizes must be carried out. Based on the colloid chemical methods, quantum dots can be fabricated in amounts of grammes like the standard fine chemicals. Above all, it is the optical properties of these quantum dots, which are interesting for their application, for example, as markers in the fluorescence microscopy of biological samples. The emitted light can be tuned throughout the whole visible spectral range up to the near-infrared range only by the variation of the particle size. Here one utilizes the fact that the distance between the discrete energy levels, relevant for optical transitions, increases with decreasing particle size. The light with the shortest wave length originates from the smallest particles (Incidentally, a similar connection exists between the resonance of the acoustic sound frequency and the spatial size of a musical instrument: the higher the note, the smaller the instrument must be).

The third path for the generation of quantum dots becomes possible because of the self-organized, spontaneous growth of well-ordered islands of uniform size in the nanometer range during the deposition of a few atomic monolayers of a semiconductor onto a substrate, under highly special conditions. These self-ordered quantum dots are the first nanostructures in the range of 10 nm, which can be produced reproducibly and also in large quantities using the standard methods of semiconductor technology. If the islands with a diameter of only a few nanometers are fabricated from a semiconductor with a small energy gap, and if they are then completely imbedded in a material with a larger energy gap, one obtains electronic quantum dots, which are well decoupled electronically from their environment. For example, quantum dots from indium-galliumarsenide ($In_xGa_{1-x}As$), imbedded into an environment of gallium-arsenide (GaAs), were intensively investigated. Also three-dimensional lattices of quantum dots can be fabricated by stacking several such layers of quantum dots on top of each other. Again, it is the opticalp properties and in particular the possibility for building a quantum-dot laser, which has stimulated the strong interest in self-ordered quantum dots. Based on these concepts, the first quantum-dot laser started to operate in 1994. Since then quantum-dot lasers have been improved considerably in terms of their quantum efficiency.

Generally, the electronic properties of quantum dots are intermediate between those of bulk semiconductors and of discrete molecules. The quantum-dot technology is particularly attractive because the active region can be modified to operate at different wavelengths by varying dot size and composition. This leads to the universal use of quantum dots in optoelectronics. In particular, it allows the fabrication of quantum-dot lasers which operate at wavelengths previously not accessible by using semiconductor laser technology. Possible applications of quantum dots include transistors, solar cells, light-emitting diodes, and diode lasers. A spectacular case is the use of quantum dots in flat-panel televisions by Sony in 2013, representing the first time they have been used in a mass-produced consumer electronics product. Quantum-dot lasers are expected to play an increasing role in optical data transmission systems.

11.6   Topological Insulators, Metals, and Semimetals

The integer and the fractional quantum-Hall-effect, which we discussed in Chap. 7, can each be understood in terms of new quantum states of the charge carriers in the two-dimensional electron gas. These quantum states appear at the edge of the two-dimensional electron gas (edge states) in the presence of a perpendicularly oriented, high magnetic field. They are robust, i.e., insensitive against geometric details and perturbations by impurities. This is the reason for the universality of the numerical values (of the electrical resistivity), which depend only on fundamental constants.

In recent years one recognized that similar quantum states with electric current flow through edge channels are possible also in the absence of magnetic fields. In electrical insulators with a suitable electronic band structure, edge states would be generated, in which charge carriers with opposite spin at a given edge move in opposite directions, different from the quantum-Hall-effect. This phenomenon is referred to as quantum-spin-Hall-effect. Presently, its experimental demonstration represents an important research subject. We will briefly sketch this development.

During the search for electrically insulating materials with an electronic band structure which is suitable for this effect, one encountered successfully the compound mercury-telluride (HgTe), a II–VI semiconductor. In this (or a similar) insulator, by means of the spin-orbit coupling, the symmetry of both spin orientations is broken. In the case of the heavy elements, the spin-orbit coupling is particularly strong, and it plays the role of an external magnetic field. The quantum-spin-Hall-effect appears at low temperatures (below 10 K) in thin layers (quantum wells), which show a sufficiently high mobility of the charge carriers, and which can be fabricated using the technique of molecular-beam epitaxy. In 2012, the three pioneers in this field, Charles L. Kane, Shoucheng Zhang, and Laurens W. Molenkamp received the highest award in the USA, for their research in the area of the physics of condensed matter, "*For the theoretical prediction and experimental observation of the quantum-spin-Hall-effect, opening the field of topological insulators*". Today, Kane teaches as a theoretical physicist at the University of Pennsylvania in Philadelphia, Pennsylvania. Zhang worked at Stanford University in California until his early death in 2018. The experimental physicist Molenkamp is heading a group at the University of Würzburg, which concentrates on molecular-beam epitaxy of II–VI semiconductors. A topological insulator is a material, which in its interior behaves like an electrical insulator, and which, at its surface (or at its edge), can transport electric charges. The topological insulators represent an advancement in the field of spintronics, which we discussed in Chap. 10 in the context with the technical applications of magnetism.

So far we discussed only two-dimensional topological insulators with their current-carrying edge states. In 2007 Charles Kane, mentioned above, together with Liang Fu, predicted that also three-dimensional forms of topological insulators would be possible, in which conducting quantum states appear at the crystal surface (surface states). A first example is $Bi_{1-x} Sb_x$, a compound of bismuth (Bi) and antimony (Sb), in which the spin-orbit coupling is particularly large. So far, the experiments have confirmed this prediction. In $Bi_{1-x} Sb_x$ the surface states are similar to the two-dimensional states in graphene. As we discussed in Sect. 11.4, in graphene near the Fermi energy, electrons and holes show a linear relation between energy and wave number, which is described by the relativistic Dirac equation of massless fermions. In two-dimensional **k**-space, the dispersion relation shows the form of two cones stacked on top of each other, the cone ends of which coincide at discrete points (Dirac points) at the Fermi energy (see Fig. 11.9). Whereas graphene shows an even number of Dirac points (namely two), in the case of $Bi_{1-x} Sb_x$ there is an odd number. As a result, in graphene the surface states are not topologically robust, and an energy gap easily opens up because of perturbations (impurities). On the other hand, in $Bi_{1-x}Sb_x$ the surface states are robust (topologically protected) and insensitive against perturbations. Shortly after the discovery of bismuth-antimony as a three-dimensional topological insulator, there were indications of topologically protected surface states also in antimony, bismuth-selenide, bismuth-telluride, and antimony-telluride.

Since the 1980s, the field of the topological materials has developed rapidly. Here the topological symmetry properties of the electronic band structure in **k**-space play an important role. In this case the highly fruitful geometric concept of the electronic band structure in k-space promoted by Lars Onsager, which we had pointed out in Sect. 5.3, becomes apparent again. Topology is a subfield of mathematics, which deals with the geometric distortion of objects and the underlying general common aspects. In a famous example it is shown, how a cup with a handle can be distorted continuously into a torus, such that both objects are topologically equivalent. The number of holes in an object remains as the important distinguishing feature. In applying these topological concepts to the electronic crystal properties, the band gap between the valence band and the conduction band takes up the role of the holes of the geometric example. In the topological band theory the possible individual cases are classified in terms of topological indices such as Berry phase and Chern number, which we do not want to discuss here any further. Band crossings between the valence band and the conduction band turned into important subjects of modern developments.

In 1996, Martin R. Zirnbauer and co-workers had developed a scheme for classification of symmetries of universal properties, which is fundamental also in the case of topological insulators. Today, Zirnbauer is director of the Institute of Theoretical Physics at the University of Cologne. In 2012 he received the Max-Planck-Medal of the German Physical Society for his work.

Topological quantum states of matter are very rare. It is noteworthy, that after more than 80 years the electronic band theory of crystals still offers surprises.

Already in 1937, the Amerikan Conyers Herring published a paper based on his PhD thesis discussing the accidental energy coincidence of individual states of the electronic band structure of the valence band and the conduction band ("contact points") in momentum space. He had performed his thesis at Princeton University under the supervision of Eugene Wigner. Herring had studied theoretically in particular the important influence of the symmetry of the crystal lattice on the stability of the contact points in the presence of impurities and other perturbations.

Recently (2015), the special cases of the structure of the conduction and valence bands linearly depending on the wave vector $\mathbf{k}$ and coinciding at discrete points in $\mathbf{k}$-space, attract increasing attention. (In Fig. 11.9 we show an example in the case of graphene.) The property of fermions having zero mass, resulting from the linear dispersion curves, leads to interesting new physical properties. These developments are based on important advances both in the experimental techniques (angle-resolved photoelectron spectroscopy, ARPES) and in the theoretical band-structure calculations (density functional theory).

Already in 1929, the German mathematician Hermann Weyl had predicted the existence of massless Fermi-particles as solutions of the Dirac-equation. Weyl had obtained his PhD in 1908 in Göttingen under the supervision of David Hilbert and subsequently worked in Zurich and in Göttingen. In 1933 he emigrated to the USA, and for many years he worked at the Institute of Advanced Studies in Princeton. There he met again Albert Einstein, whom he had known already from Zurich. The so-called Weyl-semimetals at the Weyl points in momentum space, only possess a single state, which is stable. Furthermore, at the surface of a Weyl-semimetal there appears a novel surface state. In contrast to a Weyl-semimetal, in a Dirac-semimetal the state at the Dirac-points in $\mathbf{k}$-space is twofold degenerate and instable. Presently, an intensive search is carried out for materials having these peculiarities in their electronic properties, and some success has been reported already.

11.7 Aharonov-Bohm Effect

At the end of this chapter on quantum effects in mesoscopic structures, we wish to return to the geometry of a ring in an external magnetic field. This geometry occupied us in Chap. 8 in the context of magnetic flux quantization in superconductors. Nevertheless, in the superconductor we had to deal with the macroscopic wave function of the Cooper pairs with their double elementary charge 2e. However, now we are interested in the quantum effects of the ballistic electron motion within the ring geometry of a normal conductor with sufficiently small dimensions, such that the collision processes of the electrons in the interior of the object are negligible and only the external boundary is crucial. The external magnetic field will be oriented perpendicular to the plane occupied by the ring. We assume that the diameter of the ring is much larger than the width of the ring-shaped conducting line. For the electron motion along the ring we must take into account the interference during the propagation of the matter wave along the right-half and along the left-half of the ring. It turns out that the propagation difference of the wave between the right and the left path amounts to exactly one wavelength or to an integer multiple thereof, if the magnetic field penetrating the ring area corresponds to one magnetic flux quantum (h/e) or to an integer number of magnetic flux quanta. Here it is assumed that both halves of the ring are exactly symmetrical. As a consequence of this interference between the two propagation paths we expect a periodic oscillation of the electrical resistance of the ring configuration during the variation of the external magnetic field, with the periodicity (h/e) of the enclosed magnetic flux quanta. We have been confronted with the magnetic flux quantum (h/e) before in our discussion of the fractional quantum-Hall-effect. On the other hand, if we compare the two complete trajectories around the whole ring clockwise and anti-clockwise, respectively, leading back to the same starting point, then the propagation difference between both waves after a complete revolution is exactly one wavelength, if only half a flux quantum (h/2e) occupies the ring area. Correspondingly, this results in a periodic oscillation of the electrical resistance of the ring configuration during the variation of the magnetic field with periodicity (h/2e) of the enclosed half magnetic flux quanta.

A fundamental difference between the two cases we just have discussed arises due to the fact, that the (h/e) oscillations depend sensitively upon the details of the sample. For example, the exact symmetry between the two halves of the ring becomes extremely important. On the other hand, the (h/2e) oscillations arise only from the comparison between the two complete trajectories around the whole ring along opposite directions, respectively. These two cases are related to each other by the process of time inversion. Hence, they are independent of the microscopic details of the sample. This discussed interference behavior of electron matter waves was predicted theoretically for the first time in 1959 by Yakir Aharonov and David Bohm. Hence, this is referred to as the Aharonov-Bohm effect (Fig. 11.10). At the time both scientists worked at the University of Bristol in England. In the early 1960s the effect was demonstrated in interference experiments performed with

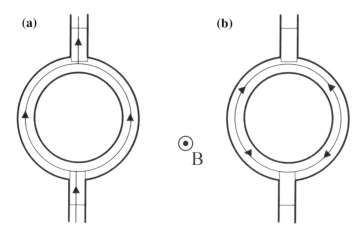

Fig. 11.10 Aharonov-Bohm effect in ring geometry. The external magnetic field is oriented perpendicular to the plane of the ring. **a** The interference between the trajectories through the right and through the left-half of the ring leads to oscillations in the electrical resistance with periodicity (h/e) of the magnetic flux enclosed by the ring. **b** For a complete revolution of two trajectories around the ring in opposite directions, respectively, the interference results in oscillations of the resistance with periodicity (h/2e)

electron beams by Gottfried Möllenstedt and co-workers at the University of Tuebingen.

For the first time the Aharonov-Bohm effect of electrons in a solid has been observed experimentally by Yurii Vasil'evich Sharvin and his son D. Yu. Sharvin in the year 1981 in Moscow. They used a thin metal cylinder made from magnesium with a diameter of 1.5–2 μm and a length of 1 cm. The cylinder was deposited as a thin layer onto a thread of quartz. The magnetic field was oriented parallel to the cylinder axis. The temperature was 1 K. The cylinder can be looked at as an object consisting of many rings stacked on top of each other. Therefore, the (sample specific) (h/e) oscillations average out, and only the (h/2e) oscillations of the electrical resistance could be observed at the time. However, the (h/e) oscillations were also detected experimentally in 1985 for the first time by the American Richard A. Webb and his co-workers at the Thomas J. Watson Research Center of IBM in the USA. They used metallic gold rings fabricated from a thin gold layer of only 38 nm thickness (Fig. 11.4). The crucial preparation step in the fabrication of the extremely small structures was the formation of a suitable mask consisting of a protecting contamination layer by means of a computer-controlled high-resolution scanning electron transmission microscope. At a temperature of 0.01 K a gold ring with 784 nm inner diameter and 41 nm width of the conducting line displayed distinct (h/e) oscillations of the electrical resistance during variation of the external magnetic field.

Chapter 12
Defects in the Crystal Lattice: Useful or Harmful?

Abstract At thermodynamic equilibrium crystals display some disorder due to the spontaneous generation of lattice defects, as we discuss in the case of lattice vacancies. Other examples of defects are color centers in ionic crystals and radiation damage in nuclear reactors. Understanding the role of dislocations in the mechanical properties of materials represented a great advance. Today, nondestructive materials testing has developed into an important field. Magnetic impurities and the Kondo effect are discussed.

During the course of many hundreds, if not thousands. of years, people have gained important and useful experience and have learned rules and recipes for the manufacture, in particular, of things made from metallic materials. At first, mechanical properties and strength under mechanical loads, exclusively dominated people's interest in materials. For example, it had been discovered early on, how long one should hammer a piece of metal in order for it to gain the optimum hardness for its use as a tool, weapon, ornament, or coin. Only in the 19th century was the cold straining and cold-work hardening systematically developed, and at the time it reached an impressively high standard, for example, in the large rolling machines of the steel industry. For a long time, this field of metallic materials was dominated by pure empiricism. The microscopic structure of wrought iron was observed for the first time only in 1863. At the time, these experimental studies were performed by Henry Clifton Sorby, who was born in a suburb of Sheffield, one of the centers of the English iron and steel industry. As an amateur geologist he was interested in the structure of rocks. After he had polished and subsequently etched his samples of wrought iron, in his light microscope he discovered characteristic structures at the sample surface, which are referred to today as the texture of a metallic sample. About 20 years later, Adolf Martens carried out pioneering research in this field, and he gained high recognition as the founder of texture microscopy and of scientific materials testing in Germany. A prominent milestone in Germany at the time was the establishment of the Kaiser-Wilhelm-Institute for Metals Research in Neubabelsberg near Berlin in the year 1920. During 1934 this Institute was moved to Stuttgart. After the Second World War the latter Institute continued in Stuttgart

© Springer Nature Switzerland AG 2019
R. P. Huebener, *Conductors, Semiconductors,*
Superconductors, Undergraduate Lecture Notes in Physics,
https://doi.org/10.1007/978-3-030-31420-0_12

as the Max Planck Institute for Metals Research. Similar Institutes were established also in the other industrialized countries. (In 2011 the Institute in Stuttgart was renamed Max Planck Institute for Intelligent Systems).

After the many discoveries in the field of electricity and magnetism in the 19th century, the electric and the magnetic material properties appeared as important new subjects, which had to be investigated. As we have discussed in Chap. 1 in conjunction with the crash of the two English Comet passenger airplanes, it is always the spectacular events and catastrophes, which impressively demonstrate the need for an almost complete understanding of material properties.

12.1 Disorder at Thermodynamic Equilibrium

Even in the purest crystal, from which all undesired impurities have been removed very carefully, there are unavoidable lattice defects for important fundamental reasons. This arises from the fact that the stable equilibrium state of a substance always requires a distinct amount of disorder. It is the "thermodynamic potential", which rules the development of the state of equilibrium in a physical system. Only in the presence of some disorder does the thermodynamic potential attain its minimum value, which guarantees equilibrium. The only exception from this exists at absolute zero temperature. The underlying ideas were developed by two physicists in the 19th century: The German Hermann von Helmholtz and the American Josiah Willard Gibbs. In this discussion of disorder, the concept of "entropy" plays a central role. The amount of disorder necessary for establishing the equilibrium can be achieved in crystals by means of the fact that, in the otherwise perfect single crystal, individual lattice sites remain unoccupied by atoms, and a certain number of lattice vacancies are generated in this way.

12.2 Vacancies in the Crystal Lattice

With the example of lattice vacancies in crystals, we will illustrate the thermodynamic treatment. We start with the free enthalpy of a crystal:

$$G = U + pV - TS \tag{12.1}$$

(U = inner energy; p = pressure; V = volume; T = temperature; S = entropy). From (12.1) we see that an increase in entropy (due to disorder) reduces free enthalpy. Because of the TS-contribution in (12.1), with increasing temperature, this reduction of free enthalpy becomes more and more important. This is exactly the reason, why in thermodynamic equilibrium there appears disorder in a crystal.

We consider a crystal consisting of N identical atoms. The presence of n lattice vacancies changes the free enthalpy G by the amount

$$\Delta G(n, p, T) = n\,U_A + n\,p\,V_A - n\,T\,S_A{}^{vibr} - T\,(N+n)S_m. \qquad (12.2)$$

In (12.2) the proportionality with n is valid only for values of n, which are small compared with N, $n \ll N$. (In (12.2): U_A = activation energy for the formation of a vacancy; V_A = activation volume for the formation of a vacancy; $S_A{}^{vibr}$ = entropy change of the lattice vibrations per vacancy; S_m = mixing entropy per particle). The mixing entropy per particle is

$$S_m = -k_B \cdot \sum_j x_j \ln\, x_j, \qquad (12.3)$$

where x_j denotes the atomic fraction of component j. In the case of vacancies, we have

$$S_m = -k_B \frac{n}{n+N} \ln \frac{n}{n+N} - k_B \frac{N}{n+N} \ln \frac{N}{n+N} \qquad (12.4a)$$

$$\approx -k_B \frac{n}{N} \ln \frac{n}{N} + k_B \ln\left(1 + \frac{n}{N}\right) \qquad (12.4b)$$

$$\approx -k_B \left(\frac{n}{N} \ln \frac{n}{N} - \frac{n}{N}\right), \qquad (12.4c)$$

(again, within the approximation $n \ll N$). In equilibrium we have

$$\left(\frac{\partial\,\Delta G}{\partial n}\right)_{p,T} = U_A + p\,V_A - T\,S_A{}^{vibr} + k_B T \ln \frac{n}{N} = 0 \qquad (12.5)$$

and

$$c(p, T) \equiv \frac{n(p, T)}{N} = \exp\left(S_A{}^{vibr}/k_B\right) \cdot \exp[-\,(U_A + p\,V_A)/k_B T]. \qquad (12.6)$$

From (12.6) we see that by plotting log c versus 1/T or versus p, one obtains straight lines (Fig. 12.1). From the slope of the plot versus 1/T one obtains $U_A + p\,V_A$ (where, in general, the contribution $p\,V_A$ at $p = 1$ at is negligible). From the slope of the plot versus p one obtains V_A. The equilibrium concentration of vacancies strongly increases with increasing temperature. The spontaneous generation of lattice vacancies contributes also to the volume expansion of the crystal, in addition to the usual thermal expansion. Therefore, the volume expansion turns out to be somewhat larger than expected only from the thermal expansion of the characteristic distance between neighbors in the crystal lattice. The volume of the vacancies must be added also. This effect was detected in a famous experiment by Ralph O. Simmons and Robert W. Balluffi. They compared the measured relative length change $\Delta L/L$ of a crystal with the relative change $\Delta a/a$ of the lattice constant,

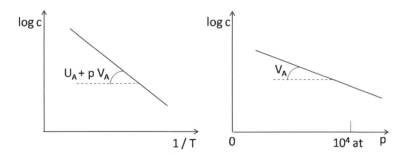

Fig. 12.1 Plot of log c versus 1/T (left) and versus p (right), schematically after (12.6)

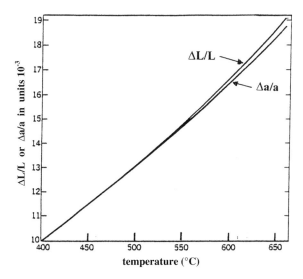

Fig. 12.2 Influence of the vacancies, existing in equilibrium within the crystal lattice, upon the temperature dependence of the crystal volume of aluminum. Because of the thermally generated vacancies, the temperature dependence of the relative length change, ΔL/L, (upper curve) is slightly larger than the temperature dependence of the relative change of the distance between neighbors in the crystal lattice, Δa/a (lower curve). The difference between both curves increases with increasing temperature. (R. O. Simmons and R. W. Balluffi)

obtained from X-ray diffraction. In Fig. 12.2 we show their results obtained with an aluminum sample.

Incidentally, the volume of a single vacancy is considerably smaller than the volume corresponding to a single atom in the unperturbed crystal (the "atomic volume"), since the neighboring atoms around the vacancy move a bit closer toward each other, and the crystal lattice becomes distorted at this location. The formation of the vacancies is accompanied by a distinct increase of the "inner energy" of the crystal. This also leads to an additional contribution to the specific heat of the crystal. In the noble metals, copper, silver, and gold, not far below their melting temperature, we have about one single vacancy per one thousand lattice atoms. In equilibrium at room temperature the vacancy concentration is smaller by many powers of ten.

Furthermore, the vacancies generated in the state of equilibrium can move through the crystal by means of diffusion. During an elementary diffusion jump from one lattice site to the next, vacancies pass through an activated state, in which they are located between two neighboring lattice sites. We denote the number of vacancies in the activated state by n^*. For the ratio n^*/n one obtains an expression analog to (12.6), in which the following quantities appear: the activation energy of motion, U_B, the activation volume of motion, V_B, and the change of the entropy of the lattice vibrations, S_B^{vibr}:

$$n^*/n = \exp\left(S_B^{vibr}/k_B\right) \cdot \exp[-\left(U_B + pV_B\right)/\ k_B T]. \qquad (12.7)$$

If the concentration of vacancies is sufficiently high, they can combine with other vacancies forming double vacancies, similar to a molecule consisting of two atoms. Still larger complexes of vacancies are also possible. In this way an extensive reaction scheme of the vacancies and their larger "molecular compounds" develops. Lattice vacancies and their motion through the crystal also represent an important mechanism for atomic materials transport in crystals and for solid state chemistry. In a crystal the process of hopping from site to site is only possible, if unoccupied lattice sites are available. Therefore, chemical reactions and the diffusion processes in a solid are closely related to the dynamics of vacancies. (This is highly important, for example, for achieving the optimum oxygen concentration in high-temperature superconductors). Before the concept of lattice vacancies had been established, one assumed that there must exist some kind of "pores in the lattice" or "loosened sites", which allow the transport of matter. In the context of this discussion it is important to note that we have only considered the case of single crystals and that, hence, we have ignored grain boundaries between single-crystalline grains with different crystallographic orientation. Very often such grain boundaries do exist, and then they provide favored diffusion channels for the transport of matter throughout the whole crystal.

Whereas in the case of the vacancies an atom is missing at its site in the crystal lattice, there is also the possibility, that one atom too much is present, which must then push itself between the other atoms and accommodate itself at an "interstitial lattice site". Again, around the interstitial atom the crystal lattice is distorted. In general, the energy gain in the crystal due to the interstitial atom is much larger than that for a vacancy, since the regular atoms in the lattice cannot be pushed away so easily, in order to make room for the newcomer. For the first time, the Russian Abram Fedorovich Ioffe proposed the idea of the interstitial lattice sites in the year 1916. During the irradiation of crystals with highly energetic particles, vacancies and interstitial lattice atoms are often generated together, if, for example, due to the particles of the radiation one atom is shot away from its regular lattice site and then must again find another place for itself within the crystal lattice. The pair of defects consisting of a vacancy and an interstitial lattice atom in the crystal is referred to as a Frenkel defect. This name originates from the Russian theoretical physicist Jakov Iljitsch Frenkel, who as a collaborator of Ioffe belonged to Ioffe's Institute in Leningrad. In the year 1925 he developed a theory of the defect pair, which was later named after him.

In the beginning of the studies of vacancies and interstitial lattice sites in a crystal, both of which are also referred to as point defects, attention was concentrated on the "ionic crystals". This type of crystal is composed of positively and negatively charged ions. Since the ions have either given up an electron or have taken up one, they possess the favored closed electron shells. Since the number of ions with the opposite electric charge, respectively, is exactly equal, charge neutrality in the crystal is maintained. The binding in the ionic crystals arises from the attractive force between ions with the opposite electric charge, as we discussed in Sect. 2.3. Ionic crystals cannot conduct an electric current and are electrical insulators. A typical example is common salt, NaCl, composed of positive sodium ions and negative chlorine ions.

Already in the 1920s, the First Institute of Physics directed by Robert Wichard Pohl at the University of Göttingen in Germany was a prominent location for the investigation of ionic crystals. In addition to his scientific research, Pohl also became famous because of the highly impressive and intuitive style of his main course in Experimental Physics, which was known as "Pohl's circus". This has also become visible in the many editions of his famous textbook on Experimental Physics consisting of three volumes. In Pohl's Institute many physical properties of the ionic crystals were studied. These crystals are transparent in a large spectral range. However, their electrical and optical properties are extremely sensitive against point defects and other perturbations in the crystal lattice. In particular, the point defects became famous, since they act as "color centers" and display characteristic optical properties. Actually, they were the first crystal defects, which were carefully studied experimentally and theoretically. Eventually, different kinds of color centers were discovered in ionic crystals, predominantly due to their optical spectral properties, and theoretical models were subsequently developed for the interpretation of the experimental results. For example, for a specific center one could show that it must correspond to a vacancy, where a negative chlorine ion (Cl^-) was missing, and where an electron was trapped. At the end of the 1930s it was mainly the Englishman Nevill Francis Mott, who applied quantum mechanics to lattice defects in crystals, similarly to the way in which it had been done before in the physics of atoms. Working in Bristol, Mott directed his attention in particular to the results obtained by Pohl's group in Göttingen. In the USA at the time it was mainly Frederick Seitz, who took up theoretical studies in this field. The point defects in ionic crystals then appeared to represent a relatively simple, but highly promising, field of study, from which valuable knowledge also about the electrical properties of semiconductors and the mechanical properties of metals could be gained.

In Göttingen, Pohl had established, perhaps worldwide, the first significant school of solid state physics. He came to Göttingen in the year 1918. Prior to that he had worked in Berlin, among other things on questions dealing with the emission of electrons from metal surfaces under light irradiation (the photo-electric effect), and at the end on problems of the radio technique. When he was asked sometime later, why in Göttingen he shifted his interest mainly to the interior of crystals, he gave the (not so serious) answer, that in the impoverished Göttingen the financial means were not sufficient for experiments carried out in high vacuum.

12.3   Materials Science of Radiation Damage

During the Second World War, in the USA new developments started because of the operation of the first nuclear reactor. On the afternoon of December 2, 1942 at about 3:30 p.m. the first nuclear chain reaction was realized in the uranium/graphite pile, which Fermi and his team had constructed below the west stand of the Stagg Field Stadium at the University of Chicago. At the time this achievement was immediately forwarded in the famous encoded announcement: "The Italian navigator has just landed in the New World. …The natives were very friendly." This event represented the start of the technical use of nuclear reactors for the production of energy. Hence, the field of defects and radiation damage in crystals and in metallic materials gained an extreme practicality. At the time, the theoretical physicist (also trained as a chemical engineer) Eugene Paul Wigner feared that the energetic neutrons generated within the reactor would cause a dangerously high concentration of lattice defects in the graphite used for slowing down the neutrons (resulting in an explosive reaction). A similar fear was expressed by Leo Szilard who, like Wigner, also originated from Hungary. The problems were then soon referred to as the "Wigner disease" or the "Szilard complication" by participating co-workers. At the time, the extreme practicality of the subject of radiation damage in crystals and in metallic materials had caused Frederick Seitz to strongly intensify his relevant theoretical calculations. Incidentally, it was Szilard, who only a few years earlier, after the discovery of nuclear fission by Otto Hahn and Fritz Strassmann in Berlin, had moved Albert Einstein to write his famous letter to the American President Franklin Delano Roosevelt, in which Einstein warned against the possibility of the Atomic Bomb.

In the years 1949–1951 Frederick Seitz established a center for basic research in the field of defects and radiation damage in solids at the University of Illinois in Urbana. Later on, from the gained knowledge one could estimate, for example, that during an operation time of 10 years of a fast breeder reactor, in its inner components each lattice atom is expelled on the average 340 times from its lattice site into an interstitial position (and back again). An early estimate for the first wall of a fusion reactor yielded a similar number, namely 170.

In the context of the doping of semiconductors we have become acquainted already with artificially generated defects or with imperfections caused by chemical admixtures in the crystal lattice. Another example are the pinning centers in superconductors which, as local perturbations of the crystal lattice, hinder the motion of the quantized magnetic flux lines, and hence, strongly reduce the heat losses during electric current flow. In both cases the defects in the crystal exercise highly useful functions. Next we will look more closely at the role of lattice defects in the mechanical strength of materials,

12.4 Mechanical Strength of Materials

In the year 1660, the Englishman Robert Hooke studied experimentally the elastic strain of metals under mechanical load, and in doing so he discovered the famous Hooke's law named after him. Later on, for a few years he was Secretary of the Royal Society in London. Hooke's law says that the elastic strain increases exactly linearly with increasing mechanical load. If the load is removed, the strain returns back to zero. The strain is still reversible. In this context the concept of mechanical stress was introduced. In the simplest case of a rod pulled in a longitudinal direction, it is the pulling force per unit cross-section of the rod. From Hooke's law the stability of metal structures can be calculated. Hence, in 1779, the first bridge worldwide made completely from iron was built near Birmingham in England. Subsequently, it has carried the road traffic for 170 years.

In subsequent discussions and in the technical use of their elastic properties, for a long time metals were treated as continuous matter, without paying attention to their inner microscopic structure. However, the still relatively simple, elastic behavior according to Hooke's law is observed only if the strain of the material does not become too large. Above a critical strain level, plastic deformation sets in, and eventually the material tears apart. Now the changes in the material due to the mechanical load are no longer reversible. At this point the microscopic structure must be taken into account. The same applies also to the changes occurring in metals during bending, rolling, or forging. However, until late into the 19th century, nobody knew what actually happened during these processes.

After the crystal structures of the metals were clarified by means of the diffraction of X-rays, the question had to be answered of how the crystal lattice is deformed during mechanical working of the metals. At the time, the largest advance in knowledge was hoped to be gained from samples consisting only of a single crystallite, referred to as single crystals. At the beginning of the 1920s at the Kaiser-Wilhelm-Institute for the Chemistry of Fibrous Materials in Berlin-Dahlem, Hermann Francis Mark, Michael Polanyi, and E. Schmid performed controlled mechanical tension tests with single crystals of zinc. Their experiments showed that the deformation of the metals under tension occurs by shifting parts of the crystal along distinct gliding planes, where the gliding plane and the gliding direction depends upon the crystal structure. During this process the microscopic crystal structure itself remains unchanged (Fig. 12.3). Experiments at the Cavendish Laboratory in Cambridge, England yielded similar results. However, during this research there appeared puzzling surprises. During their deformation, the metals seemed to become mechanically stronger. Therefore, it was presumed that, during the deformation defects are generated within the crystal lattice, which make further deformation more difficult. Furthermore, the calculation of the mechanical tension, at which parts of the crystal start shifting relative to each other, yielded values which were up to thousand times larger than the experimental data. Apparently, the metal crystals were much softer than expected theoretically. Something in the concept was wrong, and a new mechanism had to be invented.

Fig. 12.3 Deformation of a crystal due to slippage, schematically. Only the front plane of the crystal lattice is shown, and additional lattice planes are further behind. The microscopic crystal structure in the undeformed (top) and in the deformed state (bottom) remains the same. (U. Essmann)

The way out of this dilemma was provided by three scientific papers, all of which were published independently of each other in 1934. The model required had to present a strategy, in which in the final result, a more or less local defect, by its motion through the crystal lattice, achieved a gliding motion of large parts of the crystal relative to each other. In other words: a small cause must achieve a large effect. One of the authors was Michael Polanyi, who had studied the plasticity of metals for some time. The second paper was written by the Englishman, Sir Geoffrey Taylor. During the First World War, he had investigated the susceptibility of crankshafts to cracks for the Royal Air Force, and he had worked on theories about crack formation and crack propagation. Subsequently, as Royal Society Professor at the University of Cambridge, Taylor studied the plasticity of metal single crystals and the processes during their deformation. Egon Orowan, originating from Hungary, was the third author. In the 1920s he had studied electrical engineering at the Technical University in Berlin-Charlottenburg, and also developed an interest in physics. Orowan became acquainted with the problems of plastic deformation through Richard Becker, at the time just recently appointed as Professor of Theoretical Physics. Shortly before, Becker had proposed a theory on this subject. One day, Orowan had to visit Becker in his office because of a required signature. How this event abruptly changed the career of the young student, was told by Orowan later as follows:

> In the next minute my course of life was changed. This happened because of the exceptionally large office of the professor. Becker was a shy and hesitating person; however, on my way out, before I had reached the door of his huge office, he had arrived at a decision. He called me back and asked if I would not be interested to experimentally check a little theory of plasticity, which he had worked out three years ago. To engage oneself in plasticity looked like a prosaic, if not downgrading proposal during the age of a De Broglie, Heisenberg, and Schrödinger, however, it was still better than having to calculate my sixtieth transformer, and therefore, I accepted.

12.5 Dislocations

In the quoted three papers the concept of a "dislocation", which moves through the crystal as a local perturbation of the crystal lattice, was proposed for the first time. Here an additional plane of atoms is inserted into part of the crystal, which at its end within the crystal forms the "dislocation line" (Fig. 12.4). In the region around this dislocation line the crystal lattice is distorted. If the dislocation moves along its gliding plane through the crystal, at the end of this motion two parts of the crystal are displaced relative to each other by one atomic distance. During this process only individual atoms on the dislocation line are always displaced by not more than a single atomic distance (Fig. 12.5). In this way, during a deformation it is no longer necessary, to displace all atoms on the gliding plane simultaneously. In agreement with experiment, a relatively small shear stress is now sufficient to induce the motion of the dislocation. At one time, Nevill Francis Mott has vividly illustrated this process:

> The analogy with a wrinkle in a carpet is very useful... We all know that there are two methods for moving a carpet along the floor. Either we can grab one end and pull, or we can form a wrinkle at one end and drive it carefully to the other end. With a large, heavy carpet the second method needs less effort... Now we want to look at the situation in a crystal. What I have called here a wrinkle, in the technical jargon is denoted as a 'dislocation'... We see that we arrive at the same result, if a dislocation is generated at one end of the crystal and then moves through the crystal, as if one half is gliding over the other ....

The dislocation line must be understood as the boundary line of a section of the gliding plane, at which the adjoining parts of the crystal on both sides of the gliding

Fig. 12.4 Model of an edge dislocation in a simple-cubic lattice. In the lower half one can see an additional (vertical) plane of lattice atoms. (W. Sigle)

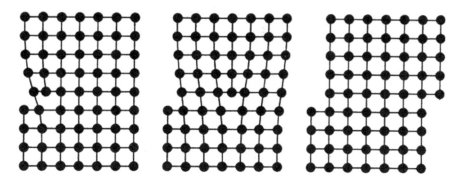

Fig. 12.5 Slippage along a gliding plane, caused by the motion of an edge dislocation, schematically. Left hand: the dislocation has been formed at the crystal edge on the left. Middle: the dislocation has reached the middle of the crystal. Right hand: the dislocation has left the crystal at the edge on the right, and has left behind a slippage step. (U. Essmann)

plane have been displaced by one atomic distance against the other. Therefore, a dislocation line cannot terminate somewhere in the middle of the crystal, and, instead, it must extend until it reaches the crystal surface, or at least it must form a closed ring. All of our discussions up to this point refers to the relatively simple case of the "edge dislocations". However, there also exist other types of dislocations, which need a more complicated description, and which will not be discussed further here.

The concept of dislocation provided the key mechanism for clarifying our understanding of the mechanical properties of crystals. At this point we recall our discussion in Sect. 8.6 of the other example where the motion of another type of defect, namely individual magnetic flux quanta, results in the key mechanism for the destruction of superconductivity and the appearance of electrical resistance.

The first direct experimental evidence for the crystal dislocations by means of their observation in an electron microscope was accomplished in the year 1956 at the Batelle Institute in Geneva and also at the Cavendish Laboratory in Cambridge. At the time, it was a particular highlight, when even the motion of dislocations could be followed in the electron microscope. The distortion of the crystal lattice near the dislocation line results in mechanical stresses in this region of the material. The stress field associated with each dislocation extends up to a relatively long distance, and by means of the stress fields an interaction arises between the dislocations. As the deformation of the crystal progresses, the number of dislocations increases. However, during this process, among the dislocations a mutual blocking effect sets in, such that the force necessary for further deformation increases. The crystal becomes mechanically stronger and harder, until it eventually breaks. The same increase in mechanical strength is achieved during the cold-working of metals, i.e., by forging, rolling, or bending. Hence, it was found essentially, that the same defect in the crystal lattice, which causes the highly useful ductility of metals, also leads to the development of their hardness during cold-working (Fig. 12.6).

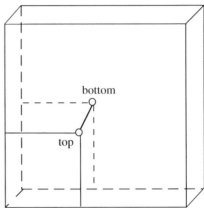

Fig. 12.6 Two-beam image showing several parallel edge dislocations in hexagonal $BaTiO_3$ ceramics. The dislocations were imaged by diffraction contrast in a transmission electron microscope. Bragg diffracting planes are bent due to the strain field of the dislocations yielding a contrast along the dislocation line. Right: sketch of a dislocation penetrating the TEM foil (about 100 nm thick) from top to bottom. (O. Eibl)

In metals, permanently changing mechanical loads are particularly harmful. There is the possibility of metal fatigue, which eventually leads to "fatigue fracture". We all know the phenomenon, i.e., that a metal wire can break, if one bends it back and forth often enough. Changing mechanical loads appear very frequently in technical equipment. As long as the load stays precisely within the elastic regime of the material, it is still harmless. However, the situation becomes critical if tiny plastic deformations start to develop, such that dislocations move back and forth within the crystal lattice. Eventually, a collection of many dislocations can lead to the seed of microcracks, representing the first stage of fatigue fracture. Here the details are highly complex, and even today they are the subject of further research. It appears that a similar scenario has led to the terrible accident of the Intercity Express Train on June 3, 1998 near Eschede north of Hannover in Germany, with many people killed or severely injured. At the time, the train ICE 884 "Wilhelm Conrad Röntgen", running between Munich and Hamburg-Altona, was involved. A hidden crack at the inner side of a metal tire eventually resulted in fatigue fracture of the tire. The tire was then caught in a switch, which considerably enhanced the disaster. The involved tire had been in operation since 1994, and it had run 1.8 million kilometers until the day of the accident. However, during these four years it was not exactly and carefully checked one single time.

12.6   Materials Testing

The accident we have just discussed raises the question of the early detection of defects in materials or of perhaps already developed microcracks or damages due to corrosion. The technical equipment required for this purpose has been available for many years, and it is being continuously improved. In the meantime, "nonde-structive materials testing"; has become an important and unavoidable technical subfield. In addition to the inspection of the raw materials of the iron and steel industry, important fields for the application of testing methods exist, for example, in armored pre-stressed concrete, in certain parts of airplanes such as wheels, fuselage, and wings, or in the under-water steel construction of derricks for oil. In the materials testing of metals a highly prominent role is played by the eddy-current method. In this method an electric high-frequency alternating current is locally induced in the test sample by means of a high-frequency coil, and simultaneously the electrical resistance behavior at this location of the test piece is determined. In this way, even very small microcracks in the interior of the material can be detected. Because of his pioneering research and development in this field over many years, which he had started in the 1930s, the German Friedrich Förster gained worldwide fame.

During recent years, also the SQUID, based on the Josephson effect and on the magnetic flux quantization in superconductors, has become increasingly important for the nondestructive materials testing and in particular also for the detection of microcracks and foreign inclusions. Having the highest sensitivity of all sensors of magnetic fields, SQUIDs are used for the detection of local anomalies in the magnetic or the electromagnetic stray field. Initially, the SQUIDs were fabricated from classical superconductors, and usually they had to be cooled down to only a few Kelvin using liquid helium. However, for a few years, SQUIDs made from high-temperature superconductors have also been available, which need to be cooled only to about 80 K, for example, with liquid nitrogen. Compared with the traditional electric eddy-current method, discussed above, the SQUID sensors are much more sensitive, in particular for the detection of defects which are located more deeply within the material. Recently, a routine evaluation at the wheel-testing facility of the German Lufthansa Airline at the Airport of Frankfurt, yielded promising results.

Using another example we wish to illustrate the importance of materials testing, in particular for constructions made of steel: i.e., the shipwreck of the *Titanic* during April 1912. Even today this marine catastrophe is a frequently-discussed dramatic subject. In 2012, exactly 100 years after the accident, it received very much attention. The physicist Uwe Essmann, working at the Max Planck Institute for Metals Research in Stuttgart, has impressively summarized this case in an essay entitled "Metals: From Stone Age Ornaments To Jet Turbine Engines", where he refers to an article in the International Herald Tribune from February 19, 1998. In the following we quote his summary:

On her maiden voyage, on April 14, 1912 shortly before midnight, the *Titanic* collided with an iceberg, and on April 15 at 2:20 a.m. she sank. In 1985 the wreck was discovered by the oceanographer Robert Ballard off Newfoundland at a depth of 3650 meters. In 1910, the hull had been assembled from steel sheets of about 2.5 cm thickness at a shipyard in Belfast, using approximately three million forged-iron rivets. By means of special robots, steel sheets from the hull and some rivets could be recovered, and it is expected that, during expeditions to the wreck in the future, additional circumstantial evidence about the course of the collision with the iceberg can be found. A group of ship-building engineers and metallurgists associated with William Garzke, Chairman of the Commision on Damages of the Society of Naval Architects and Marine Engineers, is interested in the metallurgical questions arising in this case.

Following a maneuver with the rudder, the *Titanic* hit the iceberg at its starboard side. Up to now it had been assumed that, because of this collision, a longitudinal rupture appeared across several bulkheads, which eventually resulted in the shipwreck. However, in the year 1996 investigations of the wreck with a special sonar instrument could not confirm this expectation. Instead, six lateral openings were found in the hull, which apparently were caused by blows as the *Titanic* scratched along the iceberg.

With the engineers involved in this case this observation raised the suspicion that, at the openings, the rivet seams between neighboring steel sheets had burst. Then at the (American) National Institute of Standards and Technology in Gaithersburg, Maryland, Tim Foecke started a metallurgical inspection of the salvaged rivets. By means of a dia- mond saw, one rivet was parted along its length, and its inner texture was studied with a metallurgical microscope. Forged iron does not consist of pure iron, but normally it con- tains about 2% of slag fibers, which result from slag inclusions during forging, and which are well recognized within the texture. The slag fibers improve the fatigue and the corrosion properties of the material. However, its volume fraction must not exceed 2% by much, since otherwise the mechanical strength of the rivets deteriorates. In the inspected rivet the slag content was 9%, which cannot be tolerated. If initially only a few weak rivets have given in, a seam could rupture further like in textiles, leading to a fatal influx of water. It remains to be seen, if this suspicion will be confirmed. If so, it would not change anything directly relating to the collision with the iceberg. However, if the seams between the hull sheets would have been intact, perhaps only small leaks would have appeared, which could have been handled by the bilge-pumps of the *Titanic*.

In 2008, together with Jennifer Hooper McCarthy, Tim Foecke published the book *What really sank the Titanic—New Forensic Discoveries*, in which they reported on the materials research for nearly 10 years, dealing with the objects recovered from the *Titanic* and particularly with the poor quality of the rivets.

12.7   Magnetic Impurities, Kondo Effect

As we have discussed in Sect. 5.4, in the limit of low temperatures the (nearly temperature independent) residual electric resistance of a metal is due electron scattering by structural lattice defects and chemical impurities. However, already in 1930 Walther Meissner and B. Voigt noted that in certain metals at low tempera- tures the electric resistivity displayed a minimum and then increased again by a few per cent with decreasing temperature. Subsequently, this *resistance minimum* was investigated more closely by G. J. van den Berg and coworkers. The same anomaly was observed in different metals containing a small amount of magnetic impurities (Co in Cu, for example). Although the subject was studied by an increasing number of people, it remained an unexplained mystery.

Then in 1964 the Japanese Jun Kondo entered the scene and proposed a theoretical explanation based on the scattering of electrons by magnetic impurities, which is referred to since as the Kondo effect. When Kondo considered the scattering from a magnetic ion that interacts with the spins of the conduction electrons, he made a surprising discovery. He found that one contribution in his calculation can be much larger than expected, and that the resistance of a metal can increase logarithmically at low temperatures with decreasing temperature.

Qualitatively, these results can be understood as follows. We start with the simplest model of a magnetic impurity introduced by Philip W. Anderson in 1961: There exists only one electronic level with energy ε_o below the Fermi energy occupied by an electron with its spin of $\frac{1}{2}$ pointing up (Fig. 12.7a). Quantum-mechanically, the electron can tunnel from the impurity and escape, provided its energy lies above the Fermi energy ε_F of the surrounding Fermi sea of the conduction electrons. Otherwise the electron remains trapped because all states are occupied.

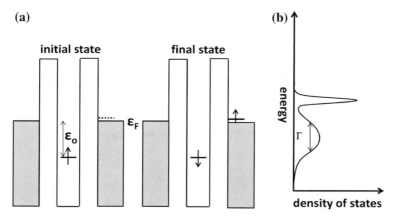

Fig. 12.7 The Anderson model describes a magnetic impurity in terms of one electronic energy level with energy ε_o below the Fermi energy occupied by an electron. Coulomb repulsion prohibits the presence of a second electron at this energy. In part **a** the potential well associated with the impurity and the Fermi sea outside the potential barrier around it is schematically indicated. Part **a** illustrates the quantum mechanical exchange process during which a spin-up electron escapes from the impurity by tunneling and reaches an unoccupied energy level above the Fermi energy. Simultaneously, an electron with the opposite spin from the Fermi sea tunnels back to the impurity site. Energetically, this exchange process becomes possible because of the Heisenberg uncertainty relation, allowing the passage through a virtual transient state during a very short time. Part **b** shows the appearance of a sharp peak in the density of states at the Fermi energy due to the presence of many magnetic impurities. Γ is the width of the energy level of the impurity, which is broadened by the short lifetime due to the electron tunneling

However, quantum mechanics allows "exchange processes", which can flip the spin of the impurity from spin up to spin down, or reverse, while simultaneously creating a spin excitation in the surrounding Fermi sea. In Fig. 12.7a we illustrate the removal of an electron from the localized impurity state and its placement into an unoccupied energy state at the surface of the Fermi sea. Classically, such a process needs an energy input of about 1–10 eV in the case of magnetic impurities. However, quantum mechanics allows such a process in form of the passage through a virtual transient state during a very short time τ because of the Heisenberg uncertainty principle. This time τ is given by

$$\tau = h/|\varepsilon_o| \tag{12.8}$$

where h is Planck's constant. Within this time τ another electron must tunnel from the Fermi sea back to the impurity. However, the spin of this electron will point in the opposite direction than before, and a spin-flip process had occurred.

Theoretically, in this case we deal again with a quantum mechanical exchange interaction. (In Sects. 10.3–10.5 we discussed the prominent role of such interaction in the case of magnetism). If the concentration of the magnetic impurities in the crystal is sufficiently small, the individual magnetic defects can be treated independently, and interference effects between them can be neglected.

This spin-exchange interaction qualitatively changes the energy spectrum of the system. The presence of many magnetic impurities results in a sharp peak in the density of states ("Kondo resonance") at the Fermi energy (Fig. 12.7b). This peak provides much additional phase space for the scattering of electrons and, hence, strongly affects the electric resistance at low temperatures.

It turns out, that below the "Kondo temperature" the electric resistance increases logarithmically with decreasing temperature. Typically, the Kondo temperature can vary between 1 and 100 K. The strong energy dependence of the density of states near the Fermi energy, seen in Fig. 12.7b, also leads to a large value of the thermoelectric power ("giant thermo-power") at low temperatures.

Nobel Prizes in Physics Closely Connected with the Physics of Solids

1901	Wilhelm Conrad Röntgen, Munich, for the discovery of the remarkable rays subsequently named after him
1909	Guglielmo Marconi, London, and Ferdinand Braun, Strassburg, for their contributions to the development of wireless telegraphy
1913	Heike Kamerlingh Onnes, Leiden, for his investigations on the properties of matter at low temperatures which lead, inter alia, to the production of liquid helium
1914	Max von Laue, Frankfort/Main, for his discovery of the diffraction of X-rays by crystals
1915	William Henry Bragg, London, and William Lawrence Bragg, Manchester, for their analysis of crystal structure by means of X-rays
1918	Max Planck, Berlin, in recognition of the services he rendered to the advancement of Physics by his discovery of energy quanta
1920	Charles Edouard Guillaume, Sèvres, in recognition of the service he has rendered to precise measurements in Physics by his discovery of anomalies in nickel steel alloys
1921	Albert Einstein, Berlin, for services to Theoretical Physics, and especially for his discovery of the law of the photoelectric effect
1923	Robert Andrews Millikan, Pasadena, California, for his work on the elementary charge of electricity and on the photo-electric effect
1924	Manne Siegbahn, Uppsala, for his discoveries and researches in the field of X-ray spectroscopy
1926	Jean Baptiste Perrin, Paris, for his work on the discontinuous structure of matter, and especially for his discovery of sedimentation equilibrium
1928	Owen Willans Richardson, London, for his work on the thermionic phenomenon and especially for his discovery of the law named after him
1929	Louis Victor de Broglie, Paris, for his discovery of the wave nature of electrons

© Springer Nature Switzerland AG 2019
R. P. Huebener, *Conductors, Semiconductors,*
Superconductors, Undergraduate Lecture Notes in Physics,
https://doi.org/10.1007/978-3-030-31420-0

1930   Venkata Raman, Calcutta, for his work on the scattering of light and for the discovery of the effect named after him

1932   Werner Heisenberg, Leipzig, for the creation of quantum mechanics, the application of which has, inter alia, led to the discovery of the allotropic forms of hydrogen

1933   Erwin Schrödinger, Berlin, and Paul Adrien Maurice Dirac, Cambridge, for the discovery of new productive forms of atomic theory

1937   Clinton Joseph Davisson, New York, N. Y., and George Paget Thomson, London, for their experimental discovery of the diffraction of electrons by crystals

1945   Wolfgang Pauli, Zurich, for the discovery of the Exclusion Principle, also called the Pauli Principle

1946   Percy Williams Bridgman, Harvard University, Massachusetts, for the invention of an apparatus to produce extremely high pressures and for discoveries he made in the field of high pressure physics

1952   Felix Bloch, Stanford University, California, and Edward Mills Purcell, Harvard University, Massachusetts, for the development of new methods for nuclear magnetic precision measurements and the discoveries in connection therewith

1954   Max Born, Edinburgh, for his fundamental research in quantum mechanics, especially for his statistical interpretation of the wave-function

1956   William Shockley, Pasadena, California, John Bardeen, Urbana, Illinois, and Walter Houser Brattain, Murray Hill, New Jersey, for their investigations on semiconductors and their discovery of the transistor effect

1961   Rudolf Ludwig Mössbauer, Munich, for his researches concerning the resonance absorption of gamma radiation and his discovery in this connection of the effect which bears his name

1962   Lew Dawidowitsch Landau, Moscow, for his pioneering theories for condensed matter, especially liquid helium

1965   Sin-itiro Tomonaga, Tokyo, Julian Seymour Schwinger, Cambridge, Massachusetts, and Richard Phillips Feynman, Pasadena, California, for their fundamental work in quantum electrodynamics, with deep-ploughing consequences for the physics of elementary particles

1970   Louis Eugène Felix Néel, Grenoble, for fundamental work and discoveries concerning antiferromagnetism and ferromagnetism which have led to important applications in solid state physics

1972   John Bardeen, Urbana, Illinois, Leon Neil Cooper, Providence, Rhode Island, and John Robert Schrieffer, Philadelphia, Pennsylvania, for their theory of superconductivity, usually called the BCS-theory

1973   Leo Esaki, Yorktown Heights, New York, and Ivar Giaever, Schenectady, New York, one half for their experimental discoveries regarding tunneling phenomena in semiconductors and superconductors, respectively, and with the other half to Brian David Josephson, Cambridge, U.K., for his theoretical predictions of the properties of a supercurrent through a tunnel barrier, in particular those phenomena which are generally known as the Josephson effects

1977 Philip Warren Anderson, Murray Hill, New Jersey, Nevill Francis Mott, Cambridge, U. K., and John Hasbrouck Van Vleck, Cambridge, Massachusetts, for their fundamental theoretical investigations of the electronic structure of magnetic and disordered systems

1978 Pyotr Leonidovich Kapitza, Moscow, for his basic inventions and discoveries in the area of low-temperature physics

1981 Kai Manne Siegbahn, Uppsala, for his contribution to the development of high-resolution electron spectroscopy

1982 Kenneth Geddes Wilson, Cornell University, New York, for his theory of critical phenomena in connection with phase transitions

1985 Klaus von Klitzing, Stuttgart, for the discovery of the quantized Hall effect

1986 Ernst Ruska, Berlin, for his fundamental work in electron optics and for the design of the first electron microscope, and the other half jointly to Gerd Binnig and Heinrich Rohrer, Zurich, for their design of the scanning tunneling microscope

1987 Johannes Georg Bednorz and Karl Alexander Müller, Zurich, for their important breakthrough in the discovery of superconductivity in ceramic materials

1991 Pierre-Gilles de Gennes, Paris, for discovering that methods developed for studying order phenomena in simple systems can be generalized to more complex forms of matter, in particular to liquid crystals and polymers

1994 Bertram Neville Brockhouse, McMaster University, Hamilton, Ontario, for the development of neutron spectroscopy, and to Clifford Glenwood Shull, Massachusetts Institute of Technology, Cambridge, Massachusetts, for the development of the neutron diffraction technique

1996 David Morris Lee, Cornell University, New York, Douglas Dean Osheroff, Stanford University, California, and Robert Coleman Richardson, Cornell University, New York, for their discovery of superfluidity in helium-3

1998 Robert Betts Laughlin, Stanford University, California, Horst Ludwig Störmer, Columbia University, New York, and Daniel Chee Tsui, Princeton University, New Jersey, for their discovery of a new form of quantum fluid with fractionally charged excitations

2000 Zhores Ivanovich Alferov, St. Petersburg, Herbert Kroemer, Santa Barbara, California, and Jack St. Clair Kilby, Dallas, Texas, for basic work on information and communication technology, in particular for developing semiconductor hetero-structures used in high-speed- and opto-electronics, and for the invention of the integrated circuit

2003 Alexei Alexeyevich Abrikosov, Argonne, Illinois, Vitaly Lazarevich Ginzburg, Moscow, and Anthony James Leggett, Urbana, Illinois, for their pioneering contributions to the theory of superconductors and superfluids

2007 Albert Louis Francois Fert, Paris, and Peter Grünberg, Jülich, for the discovery of giant magneto-resistance

2009   Willard S. Boyle, Charles Kuen Kao, and George Elwood Smith, Bell
        Laboratories, Murray Hill, New Jersey, for the invention of an imaging
        semiconductor circuit—the CCD sensor

2010   Andre Geim and Konstantin Novoselov, Manchester, U.K., for discovering
        and isolating a single free-standing atomic layer of carbon (graphene) and
        elucidating its remarkable electronic properties

2012   Serge Haroche, Collège de France and Ecole Normale Supérieure, Paris,
        France, and David J. Wineland, National Institute of Standards and
        Technology and University of Colorado, Boulder, CO, USA, for
        ground-breaking experimental methods that enable measuring and manip-
        ulation of individual quantum systems

2014   Isamu Akasaki, Meijo University, Nagoya and Nagoya University, Japan,
        Hiroshi Amano, Nagoya University, Japan, and Shuji Nakamura,
        University of California, Santa Barbara, CA, USA, for the invention of
        efficient blue light-emitting diodes which has enabled bright and
        energy-saving white light sources

2016   David J. Thouless, University of Washington, Seattle, USA, J. Michael
        Kosterlitz, Brown University, Providence, USA, and F. Duncan M.
        Haldane, Princeton University, Princeton, USA, for theoretical discoveries
        of topological phase transitions and topological phases of materials

Nobel Prizes in Chemistry Closely Connected with the Physics of Solids

1920   Walther Nernst, Berlin, in recognition of his work in thermochemistry

1936   Peter Debye, Berlin-Dahlem, for his contributions to our knowledge of molecular structure through his investigations on dipole moments and on the diffraction of X-rays and electrons in gases

1949   William Francis Giauque, Berkeley, California, for his contributions in the field of chemical thermodynamics, particularly concerning the behaviour of substances at extremely low temperatures

1954   Linus Carl Pauling, Pasadena, California, for his research into the nature of the chemical bond and its application to the elucidation of the structure of complex substances

1966   Robert Sanderson Mulliken, Chicago, Illinois, for his fundamental work concerning chemical bonds and the electronic structure of molecules by the molecular orbital method

1968   Lars Onsager, New Haven, Connecticut, for the discovery of the reciprocity relations bearing his name, which are fundamental for the thermodynamics of irreversible processes

1977   Ilya Prigogine, Brussels, for his contribution to non-equilibrium thermodynamics, particularly the theory of dissipative structures

1985   Herbert Aaron Hauptman, Buffalo, New York, and Jerome Karle, Washington, DC, for their outstanding achievements in the development of direct methods for the determination of crystal structures

1988   Johann Deisenhofer, Dallas, Texas, Robert Huber, Martinsried, and Hartmut Michel, Frankfort/Main, for the determination of the three-dimensional structure of a photosynthetic reaction centre

1991   Richard Robert Ernst, Zurich, for his contributions to the development of the methodology of high resolution nuclear magnetic resonance (NMR) spectroscopy

1996   Robert Floyd Curl, Jr., Rice University, Houston, Texas, Harold Walter Kroto, University of Sussex, and Richard Errett Smalley, Rice University, for their discovery of fullerenes

© Springer Nature Switzerland AG 2019
R. P. Huebener, *Conductors, Semiconductors, Superconductors*, Undergraduate Lecture Notes in Physics, https://doi.org/10.1007/978-3-030-31420-0

1998   Walter Kohn, Santa Barbara, California, for his development of the density-functional theory, and John Anthony Pople, Northwestern University, Evanston, Illinois, for his development of computational methods in quantum chemistry

2000   Alan Jay Heeger, Santa Barbara, California, Alan Graham MacDiarmid, Philadelphia, Pennsylvania, and Hideki Shirakawa, Tsukuba, for the discovery and development of conductive polymers

2007   Gerhard Ertl, Fritz-Haber Institut, Berlin, for his studies of chemical processes at solid surfaces

2011   Daniel Shechtman, Technion, Haifa, for the discovery of quasi-crystals

2014   Eric Betzig, Howard Hughes Medical Institute, Ashburn, VA, USA, Stefan W. Hell, Max Planck Institute for Biophysical Chemistry, Göttingen, and German Cancer Research Center, Heidelberg, Germany, and William E. Moerner, Stanford University, Stanford, CA, USA, for the development of super-resolved fluorescence Microscopy

2016   Jean-Pierre Sauvage, University of Straßburg, J, Fraser Stoddart, Northwestern University, Evanston, USA, and Bernard L. Feringa, University of Groningen, for the development of molecular machines

Bibliography

General Solid State Physics

N.W. Ashcroft, N.D. Mermin, *Solid State Physics* (Holt, Rinehart and Winston, New York, 1976)
L. Hoddeson, E. Braun, J. Teichmann, S. Weart (eds.), *Out of the Crystal Maze: Chapters from the History of Solid State Physics* (Oxford University Press, London, 1992)
Ch. Kittel, *Introduction to Solid State Physics*, 4th edn. (Wiley, New York, 1971)

Solid State Theory

A.A. Abrikosov, *Fundamentals of the Theory of Metals* (North-Holland, Amsterdam, 1988)
J. Callaway, *Quantum Theory of the Solid State, 2 volumes* (Academic Press, New York, 1974)
H. Haken, *Quantum Field Theory of Solids* (North Holland Publ. Co., Amsterdam and New York, 1976)
W.A. Harrison, *Solid State Theory* (McGraw-Hill, New York, 1969)
O. Madelung, *Introduction to Solid State Theory* (Springer, Berlin, 1978)
R.M. Martin, *Electronic Structure—Basic Theory and Practical Methods* (Cambridge University Press, Cambridge, UK, 2004)
N.F. Mott, H. Jones, *The Theory of the Properties of Metals and Alloys* (Dover Publications, New York, 1958)
R.E. Peierls, *Quantum Theory of Solids* (Oxford University Press, London, 1965)
F. Seitz, *The Modern Theory of Solids* (Mc-Graw Hill, New York, 1940)
A.H. Wilson, *The Theory of Metals* (Cambridge University Press, London, 1965)
J.M. Ziman, *Electrons and Phonons* (Oxford University Press, London, 1960)
J.M. Ziman, *Principles of the Theory of Solids* (Cambridge University Press, 1972)

Semiconductors

G. Allan, G. Bastard, N. Boccara, M. Lanoo, M. Voos (eds.), *Heterojunctions and Semiconductor Superlattices* (Springer, Berlin, 1986)
V.L. Bonch-Bruyevich, *The Electronic Theory of Heavily Doped Semiconductors* (American Elsevier Publ, New York, 1966)
K. Seeger, *Semiconductor Physics*, 8th edn. (Springer, Berlin, 2002)
B. Shklovskii, A.L. Efros, *Electronic Properties of Doped Semiconductors* (Springer, Berlin, 1984)
E. Spenke, *Electronic Semiconductors* (McGraw-Hill, New York, 1958)

© Springer Nature Switzerland AG 2019
R. P. Huebener, *Conductors, Semiconductors,*
Superconductors, Undergraduate Lecture Notes in Physics,
https://doi.org/10.1007/978-3-030-31420-0

S.M. Sze, *Physics of Semiconductor Devices*, 2nd edn. (Wiley, New York, 1969)
W. Tom Wenckebach, *Essentials of Semiconductor Physics* (Wiley, Chichester, 1999)

Superconductivity

K.H. Bennemann, J.B. Ketterson (eds.), *The Physics of Superconductors, 2 volumes* (Springer, Berlin, 2003)
W. Buckel, R. Kleiner, *Superconductivity—Fundamentals and Applications*, 2nd edn. (Wiley-VCH Verlag, Weinheim, 2004)
P.F. Dahl, *Superconductivity—Its Historical Roots and Development from Mercury to the Ceramic Oxides* (American Institute of Physics, New York, 1992)
P.G. De Gennes, *Superconductivity of Metals and Alloys* (W. A. Benjamin, New York, 1966)
R.P. Huebener, *Magnetic Flux Structures in Superconductors*, 2nd edn. (Springer, Berlin, 2001)
J.B. Ketterson, S.N. Song, *Superconductivity* (Cambridge University Press, Cambridge, 1999)
E.A. Lynton, *Superconductivity* (Methuen & Co, London, 1962)
J. Matricon, G. Waysand, *The Cold Wars: A History of Superconductivity* (Rutgers University Press, New Brunswick, 2003)
R.D. Parks (ed.), *Superconductivity, 2 volumes* (Marcel Dekker, New York, 1969)
ChP Poole Jr., H.A. Farach, R.J. Creswick, *Superconductivity* (Academic Press, New York, 1995)
R. Rickayzen, *Theory of Superconductivity* (Wiley, New York, 1965)
M. Tinkham, *Introduction to Superconductivity*, 2nd edn. (McGraw-Hill, New York, 1996)
J.R. Waldram, *Superconductivity of Metals and Cuprates* (Institute of Physics Publishing, Bristol, 1996)

Magnetism

A.S. Chakravarty, *Introduction to the Magnetic Properties of Solids* (Wiley, New York, 1980)
S. Chikazumi, *Physics of Magnetism* (Wiley, New York, 1964)
S. Chikazumi, *Physics of Ferromagnetism*, 2nd edn. (Oxford University Press, Oxford, 1997)
J. Crangle, *The Magnetic Properties of Solids* (Arnold, London, 1977)
A.H. Morrish, *The Physical Principles of Magnetism* (Wiley, New York, 1965)
R.M. White, *Quantum Theory of Magnetism* (McGraw-Hill, New York, 1970)

Special Subjects

R.D. Barnard, *Thermoelectricity in Metals and Alloys* (Taylor & Francis, London, 1972)
M. Born, K.H. Huang, *Dynamical Theory of Crystal Lattices* (Oxford University Press, Oxford, 1954)
J.H. Davies, *The Physics of Low-Dimensional Semiconductors—An Introduction* (Cambridge University Press, Cambridge, 1998)
J. Hooper McCarty, T. Foecke, *What Really Sank the Titanic* (Citadel Press Book, 2008)
K. von Klitzing, Quantum Hall Effect: discovery and application. Annu. Rev. Condens. Matter Phys. 13 (2017)
D.K.C. MacDonald, *Thermoelectricity—An Introduction to the Principles* (Wiley, New York, 1962)
A.A. Maradudin, E.W. Montroll, G.H. Weiss, *Theory of Lattice Dynamics in the Harmonic Approximation* (Academic Press, New York, 1971)
R.E. Prange, S.M. Girvin (eds.), *The Quantum Hall Effect*, 2nd edn. (Springer, New York, 1990)

A.M. Stoneham, *Theory of Defects in Solids—Electronic Structure and Defects in Insulators and Semiconductors* (Oxford University Press, Oxford, 1975)

P.T. Townsend, J.C. Kelly, *Color Centers and Imperfections in Insulators and Semiconductors* (Crane & Russack, New York, 1973)

R. Waser (ed.), *Nanoelectronics and Information Technology* (Wiley-VCH, Weinheim, 2003)

Author Index

© Springer Nature Switzerland AG 2019
R. P. Huebener, *Conductors, Semiconductors,*
Superconductors, Undergraduate Lecture Notes in Physics,
https://doi.org/10.1007/978-3-030-31420-0

239

Subject Index

A

Abrikosov flux-line lattice, 129, 131
Acceptors, 80
Acoustic modes, 42
Adiabatic demagnetization, 5, 174
Aharonov-Bohm effect, 210, 211
Anderson model, 227
Anomalous Hall effect, 75, 99
Antiferromagnetic spin waves, 182
Antiferromagnetism, 181
Artificial atoms, 205
Atomei, 8
Atomic force microscopy, 12
Atomic volume, 216

B

Ballistic motion, 195
Ballistic phonons, 45, 46
Barkhausen jumps, 178
Base, 86
Batelle institute, 223
BCS theory, 135–137
Bell Laboratories, 78, 88, 112, 141
Bicrystal, 161
Bicrystal technique, 161, 162
Bipolar transistor, 86
Bismuth-telluride, 94
Bitter decoration technique, 131, 178
Bloch ansatz, 51
Bloch frequency, 193
Bloch-Grüneisen law, 66, 67
Bloch oscillation, 192, 193
Bloch wall, 179
Bohr's magneton, 173
Boltzmann distribution, 75, 94, 171

Bose-Einstein distribution, 38, 62, 180
Bose-Einstein statistics, 40
Bosons, 40, 62
Bragg diffraction law, 22
Bragg reflection, 21, 55, 56, 65, 191, 192
Bravais lattices, 16, 17
Breeder reactor, 219
Bremsstrahlung, 26
Brillouin function, 173
Brillouin zones, 23, 25
Buckminster-fullerene, 201

C

Cadmium-selenide, 86
Carbon nanotubes, 202, 203
Cathode ray tube, 73
Cavendish laboratory, 223
Clusius separation column, 70
Coercive field, 183, 185
Coherence length, 127, 134, 156
Cold-working, 223
Collector, 86
Colloid chemistry, 206
Color centers, 218
Comet jet-aircraft, 2
Common salt, 218
Compound semiconductors, 78
Conduction band, 74
Cooper pairs, 135, 137, 140, 157, 159
Copper-oxide (CuO_2) planes, 155
Covalent bond, 32
Critical current, 120
Critical magnetic field, 120, 122
Critical state, 141
Critical temperatures, 119, 122, 155

© Springer Nature Switzerland AG 2019
R. P. Huebener, *Conductors, Semiconductors,*
Superconductors, Undergraduate Lecture Notes in Physics,
https://doi.org/10.1007/978-3-030-31420-0

Printed in the United States
By Bookmasters